Lost Spacecraft

The Search for Liberty Bell 7

Lost Spacecraft

The Search for Liberty Bell 7

by

Curt Newport

An Apogee Books Publication

Who Dares Wins.
(Special Air Service Motto)

All rights reserved under article two of the Berne Copyright Convention (1971).
We acknowledge the financial support of the Government of Canada through the Book Publishing Industry Development Program for our publishing activities.

Published by Apogee Books an imprint of Collector's Guide Publishing Inc., Box 62034, Burlington, Ontario, Canada, L7R 4K2, http://www.cgpublishing.com

Printed and bound in Canada

MISSING SPACECRAFT by Curt Newport
ISBN 1-896522-88-2 – ISSN 1496-6921

©2002 Apogee Books and Curt Newport

Contents

Introduction	7
Preface	8
Chapter 1 – Virgil I. "Gus" Grissom: *The Man, the Pilot, and the Astronaut*	12
Chapter 2 – Reaching for Space: *The Birth of Project Mercury*	22
Chapter 3 – The Mercury Spacecraft: *Building and Testing America's First Spaceship*	42
Chapter 4 – July 21, 1961: *The Flight of Liberty Bell 7*	69
Chapter 5 – Underwater Vehicles: *Pushing the Underwater Envelope*	94
Chapter 6 – Investigation: *Unlocking NASA's Archives*	109
Chapter 7 – Target Number 71: *Liberty Bell 7 Discovered*	134
Chapter 8 – Return to the Cape: *The Recovery of Liberty Bell 7*	157
Chapter 9 – Epilogue	178
Author's Note	190
Appendix A – Project Mercury Manned Flight Data	194
Appendix B – Liberty Bell 7 Technical Specifications	196
Appendix C – Project Mercury Contractors	203
Appendix D – Current Location of Mercury Spacecraft	204
Appendix E – Ocean Explorer 6000 Sonar and Magellan 725 ROV Technical Specifications	206
Appendix F – Notable Object Recovery Operations	208
Appendix G – Bibliography	209

FOR
My mother and father

And to all members of the Space Task Group,
circa 1961, Langley Research Center, Virginia

And to C. A. Lindbergh, who showed that
an ordinary man could do extraordinary things

Introduction

Tenacious and adventurous people – some of them geniuses, some of them fools – have searched the ocean floors for treasure. There has been lost gold recovered, respect has been paid to the spirits of doomed ships, and countless cannon balls, pots and pans, and bottles of very, very old wine have been brought up from the depths. Of course, many times there has been absolutely nothing recovered after months, even years of searching. The treasures are as particular as the people who seek them. In searching for his treasure, Curt Newport and his team located and recovered Liberty Bell 7, the spacecraft that carried Gus Grissom into space for the better part of fifteen minutes in 1961. Considering that the Mercury capsule defied odds as fantastic as they are documented, the news of Newport's success might have shaken the world. The Mercury capsule is not a valuable object of antiquity, though. It has the monetary worth of other Government funded vehicles from the era – say an old jeep sold as surplus or a Navy Frigate long since mothballed.

Liberty Bell 7 was a spacecraft – a marvel of science and engineering for the time. There is a classic saga of changing fortune and bitter irony in its journey from atop a rocket in Florida to the bottom of the ocean. Liberty Bell 7 did not blow up on the pad, succumb to the vacuum of space, or incinerate during its fiery battle with friction during reentry. It responded to Grissom's commands and kept him alive. Then, with a glitch as badly timed as any in the American Space program, the damn thing sank, nearly drowning the second American to fly in space. The consequences of such a horror would have been devastating to the consciousness of the world.

So history and curiosity have been served by Curt Newport finding Liberty Bell 7 and bringing it up from the bottom of the sea, making any dollar value of his discovery a mute point. The spacecraft is the equal of one of Cleopatra's barges, a ship of Columbus's fleet, or a cargo boat of Lewis and Clark. Can you imagine finding one of those crafts? Curt Newport does.

Tom Hanks
August 2002

Preface

May 2002, off Cape Hatteras, North Carolina:

Our ship is suspended at the threshold of a fast-moving river of ocean and air as I stare in disbelief at the video monitor inside the CURV III control van. Normally, the monitor shows the stable ocean surface just off our starboard rail. However, it almost looks as though we're underway, even though the satellite navigation display indicates that we're not even moving. It's early in the morning during one of our dives on a crashed US Navy twin-rotor CH-46 helicopter lost in almost 10,000 feet of water. Working from the USNS Powhatan, a Navy Fleet tug, it's our job to attach a 1¼ inch kevlar reinforced recovery line to the helo's landing gear. However, we're precariously balanced on the thin edge of a knife, supported on one side by a 40 knot sustained wind and on the other by the five knot surface current. The ROV's electrical umbilical sings in the current, creating a boiling wake as it vibrates like a guitar string.

The CURV Remotely Operated Vehicle (ROV) we're using is owned by the US Navy's Supervisor of Salvage and Diving (SUPSALV) and is a 14,000 lb. behemoth, looking more like deep diving farm equipment than the sophisticated (and expensive) tethered unmanned submersible that it is. Everything about CURV is big. The cable reel holding CURV's optical fiber umbilical weighs over 40,000 lbs. Getting the system on and off the ship takes seven tractor trailers. That's like moving a half-dozen complete households every time it goes out the door. It's almost as though the massive apparatus is beyond the ability of any mortal man to control. In addition, the logistics related to operating CURV are not trivial. Many facets of a field operation have to be dealt with including scheduling trucks, cranes, and rental cars, making hotel reservations, keeping track of the movement of ships, assigning crew berthing, as well as dealing with the numerous personalities making up any submersible operations team. From a Project Manager's standpoint, it's sort of like being a Sergeant in charge of a platoon during wartime.

Our crew is a mixture of former Oceaneering International technicians and some new people from Phoenix Marine, the current holder of the Navy's worldwide search and recovery contract. It's basically the same job but with different tee shirts. Most of them are southern Maryland boys, a few with nicknames like "Sergeant Fury, Crusty, and Superstar," who prefer burgers to Burgundy Beef, Ford trucks to Ferraris, and beer to wine. While there are no "Dirk Pitts" (author Clive Cussler's fictional hero) on our team, all of them possess a resolve to do one thing: get the job done no matter what it takes. Give these guys a few tactical nukes and the Al Qaeda would cease to exist. Of course, so would Afghanistan, Iraq, Iran, Libya, Syria, Saudi Arabia, and North Korea, for that matter. While they're basically regular guys, they do not suffer fools lightly and one mans' weakness or mistake is typically ammunition for unending ridicule. You have to be both physically and mentally tough to survive on an ROV crew: physically, to withstand the hours of back breaking manual labor; mentally, to be able to deal with numerous technical and operational problems. Also, it's important to not take things personal. You must have the ability to let things slide off you. To not do so only invites more verbal punishment.

Barry Brown, a gray-haired senior ROV technician from the other shift, pokes his head into our darkened control center saying, "Do you know that that your hydrophone pole is bent?" At first I thought he was joking. Like me, Barry's an "old timer" with ROVs and is one of the few people who has been in the business as long as I. Using small chirps of sound from the ROV, the acoustic sensor at the end of the pole tells me the location of our unmanned vehicle. And for the last hour I've been trying to figure out why CURV's location has changed position on the screen even though its been in the same spot for ages, cavorting around the upside down helicopter as it tightens up three recovery slings.

Dragging my weary body outside of the frigid confines of our computer-filled operations van, I gingerly step over assorted equipment and steel chains to check out the pole. For sure, I can see that the nine inch diameter aluminum pipe is bent and angled aft, no doubt due to the raging Gulf Stream currents. However, upon closer inspection, I discover to my horror that the pole isn't simply bent, it's broken. The current is so strong that our hydrophone pole has been snapped like a twig. It's almost like looking at a tree uprooted by a tornado. Now, the only thing holding on the lower half of the assembly is a small electrical cable and two tiny guide wires. With the help of a few of my shipmates, we haul the mangled end of the pole back on board, leaving the welding job for the day shift.

In the late morning we finally decide to recover the submersible after leaving it underwater far too long. Normally, you want to recover equipment like CURV before the weather gets too bad. But we had grown tired of sitting around and watching movies all day and were anxious to get something done. How many times can you watch *Lethal Weapon*? How long can you simply exist on a ship with nothing to look forward to except more bad weather and swift running currents? Not forever. Unfortunately, we were stuck in a situation where we were not all that anxious to recover the vehicle after all, preferring to let it roam the sea floor as long as the Powhatan could hold position two miles above. But it soon became apparent that we were cutting a few too many corners, electing in the end, to try and get our priceless underwater vehicle back in one piece. We knew that if the wind or current shifted, our unmanned submarine would be history. There's a reason that Cape Hatteras is called the "Graveyard of the Atlantic," and it's not a good one. I admit to hating the place and would be perfectly happy to never work off Hatteras again.

These days I like to take things easy. I've been working with ROV's for over two decades and when the weather's good and the ship's holding position well, I'm not in any hurry to do anything. However, this time it's different. We've got one hell of a surface current and its all the ship driver can do to keep the bow pointed in the right direction. With the CURV's recovery crane trolling for the vehicle over the starboard side, we slowly haul in the heavy umbilical, taking care not to mash anyone's fingers in the process. Incredibly, just when we need the ship to hold steady, the hand of the weather takes over, as the Powhatan is blown out of control across the surface. As our ship is pushed sideways to the wind and seas, CURV rises on top of a towering swell, threatening to recover itself for us on the wind swept deck. All we need is a few more waves like that and the vehicle will be simply plopped on board whether we like it or not. The crane operator tightens our high-tech Spectra recovery line at the last moment, plucking the seven ton contraption out of the ocean. As we cower on deck, a strong blast of wind drenches us with buckets of sea water from the sub, while we almost blindly reach out to steady the swinging submersible. Guided by a few sets of hands, CURV swings to a stop onto the rusting steel deck and all is over. We've somehow managed to dodge another bullet.

A few days later, after refueling our ship in Portsmouth, Virginia, we head back out in our ship, the world's most expensive movie theater, as we wait for the weather to abate and the current to subside. However, it soon becomes obvious that the sea conditions are what they are and if we cannot deal with them, we might as well pack up our gear and go home empty-handed. But we do not want to do this. Instead, we decide to modify our normal recovery procedure and go for broke. Typically, on a Navy deep water recovery, we would lower a massive aluminum spooler holding our liftline to the bottom near the crashed helicopter and attach the end of the line to the helicopter. Next, we would use CURV to haul the spooler up to the surface, deploying the line in the process. However, the currents ripping through the area make it doubtful that we can get the spooler close enough to the helicopter, mostly because our navigation system has not been all that accurate. Instead, we reluctantly decide to remove all 14,000 feet of lift line (almost three miles) from the spooler on deck and attach it to the CURV's electrical umbilical with duct tape and plastic tie-wraps. This way we can haul the line directly to the submerged helicopter with CURV. Removing this much line in this manner is no easy task, to say the least, as it means that we have to manually rotate the eight foot diameter spooler. We take turns at this physically exhausting work; using your feet to rotate this massive ball of string is like working out on the Stairmaster from Hell for 12 hours nonstop. With CURV underwater the whole time, this work takes almost two days.

Finally, we connect the line to the helicopter, recover the vehicle, and watch as the 25,000 lb. aircraft is dragged out of the water on to the deck of a second ship, one with sufficient deck space to hold the mangled remains of the totaled Boeing / Vertol machine. We're done. We head back to port, anxious to get acquainted with the soft beds of a nearby hotel, a bathroom that doesn't reek, and the simple silence of a room not invaded by the rumbling sounds of a ship's bow thruster. Our crew drinks. Our crew sleeps. Our crew eats a decent and unencumbered meal. We do not watch any more movies.

While during past operations I normally piloted the submersible, on this mission I was mostly monitoring the underwater navigation system, which is fine with me. It's an easy job with equipment I know well enough and I no longer have anything to prove to anyone. Why? Because I proved it on July 20, 1999. In the early morning hours of that day, I led a team that recovered the Liberty Bell 7 Mercury spacecraft from over 16,000 feet of water in the Atlantic Ocean. From its beginnings as an underwater pipe dream to reality, the effort took 14 years. If I was an actor, recovering Liberty Bell 7 was like winning an Oscar. If I was a professional racing driver, I had just won the Indy 500. How can you beat that? The truth is, you can't. How could an idea that was so farfetched actually be accomplished? How could an unknown like me do something that many people said was impossible? This is how one lost spacecraft was found and recovered from the Atlantic Ocean.

Chapter 1 – Virgil I. "Gus" Grissom
The Man, the Pilot, and the Astronaut

There are no signposts in the sky to show a man has passed that way before. There are no channels marked. The flier breaks each second into new uncharted seas.
— Anne Morrow Lindbergh

When the writers of the popular CBS television program "CSI" needed a last name for actor William Peterson's character, I don't think it was an accident they chose "Grissom." The name alone suggests a quiet professionalism and dedication to the job at hand; one of the most obvious traits of CSI's "Gil Grissom." Why? Because that's what Gus Grissom was all about. Gus didn't have (and didn't want to have) the charisma of a Wally Schirra or the "fame" of a John Glenn; for the most part, he studied his craft of flying and simply made every effort to do the best job possible.

Other than Virgil Ivan Grissom, only two other people of any notoriety have ever been created by the town of Mitchell, Indiana: actor Claude Aikens and Sam Bass, a notorious Kansas bank robber. In 1910, the population of Mitchell was about 3,500 people; eighty years later during the 1990 census, only 1,169 additional people thought Mitchell a place worth relocating to. Clearly, unless a miracle happens, Mitchell is one of those small Midwestern towns on the fast track to oblivion. But for Gus, Mitchell was a good place to grow up. The people were friendly, instilled small-town values, and worked hard for what they had.

The thing to remember about Mitchell is that from a teenager's perspective, in the 1940's there wasn't much excitement in town. Gas rationing was the rule and parents needed those few gallons per week to get back and forth to work, which left little for having any fun. If someone wanted to do anything with their lives they had to get out of Mitchell and that's what Gus did.
— Bill Head, Grissom's Childhood Friend

Growing up in Mitchell had to be similar to living in Enterprise, Alabama, a small town I spent some years in during the late 1950's while my father was stationed at Fort Rucker; I remember walking up a hill near our house to buy a small carton of chocolate milk off the loading dock of a dairy for a nickel. Gus was born at 8:00 p.m. to Mr. and Mrs. Dennis Grissom on April 3,

1926, to later be the oldest of four children, his brothers Norman and Lowell and his sister Wilma. After moving to Mitchell from nearby Martin County in 1924, Dennis Grissom found a good job working for the Baltimore and Ohio Railroad, a company he worked for his entire life. Norman later found work as a linotype operator for the Mitchell Tribune and Lowell ended up working as a systems engineer for McDonnell Aircraft, the same company that created Liberty Bell 7.

As a youth, Gus was fascinated by the sounds and appearance of power, such as the steam engines running along the rail line situated only a block from the Grissom home. He was interested in popular science, in particular aircraft technology, and building model airplanes. As a student in grade school, high school, and college, Grissom received average grades, except for those subjects that caught his interest; in those, such as math, he excelled. Gus' first grade school teacher, Mrs. Myrtle McKeever, remembers Virgil as being shy and retiring. However, she also recalled that he was an exceptionally good student with a special interest in numbers. A harbinger of Grissom the test pilot and engineer? Perhaps. Regardless of Grissom's performance as a student, he was pretty damn intelligent and had an IQ of 145, which puts him in the gifted range. As with many individuals, Gus probably did enough to get by on subjects he had no special interest in. But when it came to something he could dig his teeth into, he devoured it with a resolve that would be his most notable trait as an astronaut.

At high school age, Virgil Ivan Grissom's frame stood only 5 feet 4 inches tall, weighing slightly less than 100 lbs. Even though he was not large enough for varsity sports, he was very athletic, played basketball, and loved to swim. Even more important was that he had excellent eye, hand, and foot coordination and was very competitive. In other words, what he lacked in size and mass, he made up for by trying harder. From the human factors standpoint, it meant that he had the ability to quickly and accurately command his body to do what he wanted based on what he observed with his eyes. Everyone can do this, but the people who make skilled pilots are really good at it, having the ability to control aileron, rudder, and elevator simultaneously, track an enemy aircraft, and lead them just the right amount before squeezing off a hail of gun fire.

Overall, Gus had the ability to make quick decisions based on what he saw, do things with uncanny calmness, and exhibit a fierce determination to prevail against all obstacles – the perfect combination of abilities for a fighter pilot. In addition, by his selection as an astronaut by NASA and due to

Grissom's small stature, McDonnell Aircraft engineers got an automatic weight savings of about 40 lbs. over all of the other Mercury astronauts.

> *Whenever I've started out with something new, I've been concerned that I won't perform with the big boys. I'm always surprised to find that I can do as well as anyone else. I know I'm going to be scared when I get in the capsule, but I don't worry about being scared... I won't be scared very long. ... I know it will work.*
>
> —Virgil I. "Gus" Grissom

Grissom, along with two of his friends, were inducted into the Air Force right after high school in August, 1944. While he hoped to become a pilot, after going through basic training at Sheppard Air Force Base, he was instead sent to Brooks Field in San Antonio, Texas, stuck behind a desk as a clerk typist. He hated it. However, he got married that year to Miss Betty Lavonne Moore who, like Grissom, was a Mitchell native. By then, Gus was stationed at Boca Raton Air Force Base; he was discharged in November of 1945 holding the rank of Corporal.

After Gus returned back to Mitchell, he became more and more restless; he wanted to fly. He also knew damn well that without a college degree, he had little chance of becoming a pilot. As a result, in September of 1946, Gus managed to enroll in Purdue University in their mechanical engineering program. "Managed" is the right word because Gus did not come from Mitchell with glowing recommendations from his former high school. It wasn't that he was a bad student, just that he and his best friend were an explosive combination.

Grissom had developed a close friendship with Bill Head, another Mitchell resident, during high school and his studies at Purdue University. As Bill remembers it, "We were both not that big and always seemed to be grouped together in school . . . we also had a lot in common which helped us become close friends." In fact, this friendship almost cost Grissom a spot at Purdue. When he applied for college, the Mitchell High School principal did not give him a particularly good recommendation, the college counselor commenting (of Grissom and Head) that, ". . . looks like you two boys sat too close together in school." It was also during his work at Purdue that Grissom got the nickname "Gus." Apparently, one of the other students thought his name *was* Gus and kept calling him by that name. It stuck for some reason and when Virgil Ivan Grissom was selected as a Mercury astronaut, "Gus" was chosen over Virgil (which did not sound like an "American" name) and Ivan (definitely too Russian for during the Cold War).

Even though Grissom was attending college on the G.I. Bill, he and Betty still needed money to survive. His new bride found work as a long-distance telephone operator and Gus flipped hamburgers at a local restaurant. Poverty.

He was a better than average student and was a very determined young man who wanted more than anything else in the world to become a test pilot.
— Professor David D. Clark, Purdue University

Even though Grissom ran out of money on his G.I. Bill a year before he graduated, he finished his courses a semester earlier while carrying a very heavy course load (and working 30 hours a week in the restaurant at the same time). Unlike the other younger college students, the older and more mature Grissom didn't fool around when it came to college. He knew what he wanted, plotted his strategy, and got his degree. Grissom the man, was slowly but surely being transformed into Grissom the test pilot.

The future Liberty Bell 7 driver rejoined the Air Force in 1950 after being assigned to the Air Material Command, then the organization having responsibility for test pilots. From then on, he was shuffled to various installations such as Randolph and Williams Air Force Bases, ending up in Arizona for advanced flight training. When Gus' first son, Scott, was born that same year, Betty was stuck with the newborn in Indiana as neither of them had enough money to make trips to Arizona or Indiana. Finally, later on in the year, Betty and Scott made it to Arizona where the new family rented a small apartment near the base. But it was not to last. The Korean conflict was starting to boil in the Far East and it was not long before Grissom the pilot trainee became a fighter pilot.

Grissom was sent to Korea in December of 1951 and assigned to the 334th Fighter Interceptor Squadron, based at Kimpo Air Force Base, only 12 miles from the front lines. In 1951, flying air combat in an F-86 Sabre jet was not a whole lot different than shooting it out in the Second World War. The big difference was that the closure speeds were phenomenal (over 1,000 mph in a head-on contest), i.e., there was a lot less time to react. Gus' F-86 was named "Scotty," after his firstborn, and it didn't take Grissom long to get into the fray.

Gus, a taciturn, grizzled fellow, had flown a hundred combat missions in Korea. One of the stories told about him was how, when he first got to Korea, he found that pilots who had not been shot at by a MiG weren't allowed a

> seat on the bus to the hanger. Gus stood only once. He had shot it out with a MiG on his first mission to qualify for a seat — and the "brotherhood of the right stuff"... he filled the bill as the prototypical test pilot. He could sit at a bar for hours, and he never failed to notice a pretty girl in the room. He could be cranky and tough, but he went his own way. He wasn't a hanger on. He was a hard liver and loved to party, but if Roger Chaffee, the youngest astronaut in the program, was pulling some notably boring duty, you were likely to find Gus sharing it with him.
>
> — Walter Cunningham, Apollo Astronaut

Grissom's primary adversary in Korea was the MiG-15 fighter, a worthy opponent armed with both 23 mm and 37 mm cannons. Gus' tool was the F-86 Sabre, probably the E model, which were delivered to his squadron in 1951 (they did not receive the F model until July of 1952, probably after Grissom had already served his tour of duty). Flying an F-86 in Korea was akin to being a gunfighter armed with six .50 caliber machine guns. In fact, the first time Grissom was shot at, he didn't even know what was happening, "I was flying along up there and it was kind of strange. For a moment I couldn't figure out what those little red things were going by... then I realized I was being shot at," the pilot later remembered. Unlike the modern jet fighter aircraft of today, you only shot what you could see. Grissom did well during his tour, being selected to fly wing for the squadron commander. This was an important position and it said something about Gus' flying abilities: his boss thought he was good enough to protect his back in the dangerous skies over Korea. However, it also meant that Grissom spent more time looking after someone else than trying to kill MiGs himself. As a result, Gus never shot down a MiG. He did well nonetheless:

> On March 23, 1952, Lieutenant Grissom was leading a flight of F-86's flying top cover for a photo reconnaissance escort mission over Korea. Upon completing its mission, the reconnaissance aircraft was attacked as it left the target area. Lieutenant Grissom dispatched two of his flight to counterattack, and the two MiG-15's were completely routed. During this encounter, two additional MiG-15's dived in to assist the initial attack but Lieutenant Grissom skillfully led his flight to intercept this new threat... The superlative airmanship demonstrated by Lieutenant Grissom on this mission exemplified his tour of duty, reflecting great credit upon himself and his comrades in arms of the United Nations and the U.S. Air Force.
>
> — Grissom's Citation for the Distinguished Flying Cross

In general terms, what Grissom did was successfully lead his flight of fighters into an attack on several MiG-15's, managing to get into a firing position several times, and probably scaring the hell out of the North Korean pilots.

> *We chased the MiGs around, and the MiGs chased us around, and I usually got shot at more than I got to shoot at them. I decided that space flight could not be more dangerous than that. You get used to handling yourself in a situation like this, when death is supposedly knocking at your door. You are scared, but you learn to take care of yourself... if they would have let us fly over the Yalu River I probably would have been an Ace.*
> —Virgil I. Grissom

Flying 100 combat missions in six months means that Grissom flew about every other day, racking up hundreds of hours in his F-86 (his request for 25 additional missions was denied by the Air Force). It also meant that the pilot who returned to Bryan Air Force Base in Texas was not the same individual who left the United States half a year earlier; flying in combat and being shot at changes a person. They can become more fatalistic, even callous, in some cases when it comes to the value of human life. As with many combat pilots, Gus never talked a whole lot about what he did in Korea, at least not to his brothers or Bill Head. He also left Korea with the Distinguished Flying Cross, the Air Medal, and the knowledge that neither he nor any of the leaders he flew wing for were ever hit by MiG gunfire. Another thing is certain; Grissom developed flying skills in Korea and character traits not obtainable any other way. He had to have been more confident in his abilities as a pilot and as a man.

Gus' job at Bryan was as an instructor in jet flying and that can sometimes be more dangerous than combat, as Grissom later recalled; "I know what I'm going to do up there all the time... but I don't know what that student is going to do. At least in combat you have a general idea as to what the other pilot is trying to do – kill you." Students can be unpredictable and on at least one occasion Grissom and his student almost drilled a hole into the ground with their two-seat jet trainer. However, being a flight instructor also makes a person a better pilot as they see firsthand the mistakes of others and learn from them. Grissom watched and learned.

Mark Grissom, Gus' second son was born in 1954 and that same year the pilot enrolled in the Institute of Technology at Wright Patterson Air Force Base in Dayton, Ohio. He was finally where he really wanted to be: on the cutting edge of jet technology and in training to become a test pilot.

> *What I remember most of all about Gus was the thoroughness with which he approached everything he did; and this carried over into things not related to flying. Gus also had a dry wit of sorts. When one of the Air Force's Texas Towers collapsed in the Atlantic Ocean with a loss of all aboard, someone commented that they were going to have a hard time finding people to go back out on the things. Gus remarked, 'all they have to do is cut the orders . . .'*
> — Lowell Grissom, Gus' Younger Brother

In the late 1950's, the future astronaut was now in the middle of what could be called one of the golden ages of aviation. Grissom was flying on the spearhead of a new era of aircraft technology, much like Charles Lindbergh, Jimmy Doolittle, Bob Hoover, and Chuck Yeager had done years before. At Wright Patterson, Gus quickly developed a reputation as "one of the best jet jockeys in the business," as he racked up the flying hours in the Air Force's latest aerospace technology such as the F-104 Starfighter, a 1,400 mile per hour "missile with a man inside." The silver jet, which had wings only 7½ feet long, was also called by another name: the *Widowmaker*. It was also during this stint that he met and became friends with another future Mercury astronaut, L. Gordon Cooper. While Grissom tested new fighter aircraft, he also honed his engineering skills, studying aeronautical engineering and experimental test flying.

> *Grissom was good. We'd heard that he was a pilot's pilot, a talented engineer, and easy to like.*
> — Christopher C. Kraft, NASA Flight Director

Test piloting is a different kind of flying and it doesn't necessarily involve the wild aerial maneuvers depicted in the movies. What Grissom did was to determine the capabilities of previously unproven jet aircraft by flying them in a very structured manner in various attitudes while recording what happened. Air Force regulations stated at the time that the pilot, "plans, coordinates, and conducts flight test procedures on experimental and production type aircraft to evaluate and report the flight characteristics, performance, stability, and functional utility as a military weapons system." In addition, test pilots have to view the performance of aircraft not only from their own standpoint and abilities, but from the perspective of the average Air Force pilot who has not had the benefit of their advanced training.

The four years Grissom spent in Dayton were enjoyable as he was able to lead somewhat of a normal life. Rather than being stuck on some far-away military base away from his family, he spent the evenings at home with Betty, Scott, and Mark and had regular work hours. This lifestyle was the exception as opposed to the norm. A military family by definition has a very transient existence with the wife and children following the husband all over the globe from one billet to another. It's particularly difficult for the wife, and Betty faced weeks and months alone while Gus was off fighting wars or testing some new aircraft. She had to pay all the bills, clean the house, make sure the kids got fed and off to school, all while being the "perfect military wife." Also, being transferred from one military base to another was probably tough on Scott and Mark. As soon as they made friends, it was off to another home and school. They were always the new kids in class, having to be paraded in front of their classmates every few years and introduced as "new students." It could be a lonely existence.

Grissom finished up his studies and attended the test pilot school at Edwards Air Force Base in Nevada, where in 1957 he received his credentials and found himself back where he started: Wright Patterson Air Force base. "This was what I wanted all along, and when I finished my studies and began the job of testing jet aircraft, well, there wasn't a happier pilot in the Air Force," Grissom recalled. However, it was not long before Grissom received a Top Secret teletype message which would change his life.

Unbeknownst to Grissom, a new government agency called the National Aeronautics and Space Administration (NASA) was seeking skilled test pilots for a new program which didn't even have a name. On December 22, 1958, NASA Project A, announcement No. 1, was an invitation for qualified individuals to apply for a civil service position of "Research Astronaut-Candidate." Tentatively, it was called "Project Astronaut." It would later be called Project Mercury. Although NASA was issuing public requests for personnel, in keeping with President Eisenhower's wishes, they were also sending out orders to qualified military test pilots to report to Washington for an interview. Grissom was a logical candidate.

> . . . he [the adjutant at Wright Patterson] handed me a teletype message form, classified 'Top Secret.' It said that Captain Grissom would report to an

address in Washington, D.C., by such and such an hour on such and such a date. What really intrigued me was the order that I should wear civilian clothing. On the appointed day, wearing my best civilian suit, and still as baffled as ever, I turned up at the Washington address I'd been given.
— Virgil I. Grissom

Surprisingly, when Grissom returned home that day and told Betty about the message, she replied, "What are they going to do? Shoot you up in the nose cone of an Atlas?" Betty had no idea how right she was.

By that day in 1958, Grissom had logged over 4,600 hours as a pilot, 3,500 of which had been in jet aircraft. Overall, 69 men received teletype messages ordering them to report to Washington. Out of this first group, 56 pilots actually took the written tests, technical interviews, psychiatric and medical examinations. By March, there were 32 men left in this NASA pool of raw aeronautical talent.

In examining the men, all of them near-perfect human specimens, NASA had to try and figure out who would make the best astronauts. However, the closest anyone had come to actually being in space was while piloting the assorted high-performance aircraft at Edwards. As a result, the actual selection criteria were dictated not by one person, but by a committee consisting of a combination of a test pilot / engineers, two flight surgeons, two psychologists, and two psychiatrists. These were the people who had interviewed and studied the 69 men in Washington before herding them off like cattle to the doctors at the Lovelace Foundation for Medical Research in Albuquerque, New Mexico. The men who were *not* one of the 18 finalists were assured that medical data from any of the specialized tests would never find its way into their service records. They were truly the best, if not among the best America had to offer in the way of steely-eyed test pilots. And they were to be *pilots*, not guinea pigs. Scott Crossfield, the famed test pilot put it best saying, "Where else would you get a nonlinear computer weighing only 160 pounds, having a billion binary decision elements, that can be mass produced by unskilled labor?"

Grissom passed all of the tests with no problems, except for the one dealing with allergic sensitivities; it was then that the pilot discovered that he was susceptible to hay fever. He managed to talk the doctors out of disqualifying him, saying, "There won't be any ragweed pollen in space."

Grissom had reservations about NASA and Project Mercury. He was doing fine in the Air Force and working with a talented group of people. Not only that, he was stationed at Wright-Patterson and having a blast flying the latest examples of aircraft technology. What would happen if he didn't like NASA and wanted to return to the Air Force? Would he lose all of his seniority? Would his flying career be destroyed? Another thing was that the project sounded more like a circus stunt as opposed to "real" flying. There might not be any actual flying involved at all, if the capsule was automated. Still, the idea of going higher and faster than any of his fellow pilots intrigued him, as did the engineering aspects of the project:

> We all like to be respected in our fields. I happened to be a career officer in the military – and, I think, a deeply patriotic one. If my country decided that I was one of the better qualified people for this new mission, then I was proud and happy to help out. I guess there was also a spirit of pioneering and adventure involved in the decision. As I told a friend of mine once who asked me why I joined Mercury, I think if I had been alive 150 years ago I might have wanted to go out and help open up the West.
> — Virgil I. Grissom

Even after he was accepted into Project Mercury, Grissom had second thoughts at times about space flight, later remarking, "After I had made the grade, I would lie in bed once in a while at night and think of the capsule and booster and ask myself, 'Now what in hell do you want to get up on that thing for?'" However, Grissom could not resist the exciting opportunity to do something that he had never done before: pilot a real spacecraft.

Chapter 2 – Reaching for Space
The Birth of Project Mercury

Space isn't remote at all. It's only an hour's drive away if your car could go straight upwards.
— Sir Fred Hoyle

When Grissom finally decided to join Project Mercury, much had already been done towards the goal of sending a man into space. For one thing, even before the future astronaut knew what it was, the National Aeronautics and Space Administration, or NASA, had been created by President Eisenhower. What he did was take the National Advisory Committee for Aeronautics (NACA) and use it as the basis for a new civilian agency, which was activated on October 1, 1958 and headed up by Dr. Keith Glennan. However, that was only the beginning of NASA, and long before there was an official space agency, before Grissom became an astronaut, and before there was a Mercury spacecraft, people thought about the idea of space travel.

The idea of sending someone into space was a natural extension of mans' curiosity about the world and what lay beyond (also irrevocably tied to establishing territories and exploiting technological capabilities). The Romans knew how to build roads and aqueducts, allowing them to transport armies and materials and supply their cities with fresh water. The Vikings and Spanish were prominent because they had ships which could be used to wage wars and establish new settlements in locations like Greenland and the New World. In more recent years, any country with a head start in aviation technology had the opportunity to project force using aircraft to influence events over a wide geographic area. In other words, rather than restricting their capabilities to the Earth's surface, a country could establish a hold on the upper atmosphere, and possibly even space itself: the ultimate high ground. From the strategic and political standpoints, having the ability to travel through space is about control, as in using submarines to govern areas underwater, aircraft to control the atmosphere, ships to rule the ocean surface, and armies to roam the ground.

The Chinese experimented with rockets in warfare as early as 1232 A.D during a siege laid to the city of Kai-fung-fu by the Mongrels. They had "arrows of flying fire," which these days is taken to mean rockets. However, the practical development of modern rockets was really influenced by Dr. Robert Goddard and the German Rocket Society (the Verein für Raumschiffahrt or VfR).

Robert Hutchins Goddard was a well respected physics professor from Clark University who in 1919 wrote a pivotal scientific paper describing the prospects for space travel entitled, "A Method of Reaching Extreme Altitudes." While the publication was really a professional study of certain phases of physics, one idea suggested that a rocket be built which could hit the Moon and explode a charge of flash powder. This simple concept startled the local press, mainly because Goddard was not a crackpot, but a well known university professor. At the time, it was a fantastic idea that people had difficulty comprehending because the general public could not imagine how the Moon could be hit with anything, even a rocket. In fact, Arthur C. Clark once quoted a man who had written to him as late as the 1950's to inform him there was a barrier separating the outer atmosphere from space proper, an "adamantine membrane" which kept our air in. Clearly, there were many people who could not comprehend the idea of travel outside of the Earth's boundaries – mostly, because in the early 1900's, rockets were things you fired on the Fourth of July, not something you launched into space.

After being sent to the Smithsonian Institution, the paper was closely studied by scientists in both the United States and – Germany.

In 1926, Goddard launched the first liquid-fueled rocket from a farm in Auburn, Massachusetts, the primitive vehicle reaching an altitude of 184 feet in 2.5 seconds. This came after he succeeded in operating a small rocket motor fueled by liquid oxygen and gasoline. The event was described in Goddard's second Smithsonian report. However, by 1931, the professor's activities had outgrown the confines of Massachusetts and with support from the Guggenheim Foundation, Goddard relocated his work to New Mexico near what would eventually become the White Sands Missile Range. Goddard's scientific work had earlier come to the attention of famed aviator Charles Lindbergh, who was instrumental in helping the reclusive scientist receive much needed financial support.

By now Goddard's accomplishments in New Mexico were beginning to come to the attention of both the American and German scientific community, in particular a German mathematics professor named Hermann Oberth. One of Oberth's associates, Willy Ley, wrote to Goddard asking about reprints of his technical papers:

Please allow me to extend my congratulations to the successful rocket flight you just performed. Naturally I am anxious to learn more about the flight and from my own experience I know it is highly insufficient (to say the least), to depend upon newspaper reports. I hope you will publish more about your rocket shot... and I would highly appreciate it if you would be good enough to inform me in which journal ... your publication will appear...
— Will Ley, January, 1936

Oberth was definitely a major player in early rocket development and, like Goddard, published his own scientific paper in 1923 about the possibilities of space travel entitled, "Die Rakete zu den Planetenräumen" (The Rocket into Interplanetary Space). While the book was highly technical and meaningless except to other scientists, it was a limited popular success. As Willy Ley later put it, ". . . it [Oberth's book] sought the professional criticism of the professionals – but, alas, there was no such profession." In 1927, Oberth was asked to join the VfR and was later hired as a scientific advisor to movie director Fritz Lang for a film entitled "Frau im Mond" (The Girl in the Moon).

In 1932, one of the members of the VfR was a young engineering student named Wernher von Braun, later to become the director of the Marshall Space Flight Center. As Hitler came to power in Germany, von Braun managed to get himself a job as a civilian employee of the German Army, it then being responsible for the development of liquid fuel rockets. This effort would eventually result in the creation of several rockets, among them the A-1, a 4.6 foot design with a rotating center section, the A-3, a 21 foot tall creation with a new 3,300 lb. thrust motor, and the A-4, later to be renamed by Hitler as the V-2 (Vergeltungswaffe Zwei, or Revenge Weapon 2). The larger weapons were tested at Germany's new Peenemünde facility, where they could be aimed towards the Baltic. One reason Germany developed this new technology was that rockets, as opposed to artillery, were not regulated by the Treaty of Versailles, allowing the formerly beaten nation to exploit a loophole in the treaty. The first operational use of the 46 foot tall, 8,837 lb. (empty weight) V-2, which carried a one ton warhead of Amatol, was in 1944 when two rounds were fired against Paris. When fully loaded, the fuel tanks of the first ballistic missile accounted for three-quarters of the weight of the rocket.

There were a lot of similarities between Germany's V-2 rocket and the Redstone booster used during Grissom's Liberty Bell 7 mission years later. This was not surprising as the Redstone was designed by ex-Peenemünde engineers working under Dr. Wernher von Braun. First of all, both missiles used alcohol for fuel and a liquid oxygen oxidizer, fed by turbo pumps to a fixed (i.e., not gimbaled) combustion chamber and exhaust nozzle (the direction of thrust was controlled using graphite vanes mounted underneath the exhaust nozzle). Secondly, both the V-2 and Redstone had external vanes on their fins which were used to control the direction of flight through the atmosphere. The internal vanes helped during the critical early stages of flight until the missile was going fast enough for the external fins to have an effect. However, once the rocket had reached a high enough altitude, the external vanes were useless and it was up to the graphite internal vanes again to keep everything pointed in the right direction. Overall, while Grissom's Redstone was taller and carried more fuel, the basic design was very similar to that of a tactical missile developed by the German Army over 15 years earlier. The V-2 had an instrument section where its flight path could be pre-programmed into each missile before launch. All the ground crew had to do was make sure that the number 1 and 3 fins were pointed at the target because the tilt program was referenced to those fins on the missile. However, once a V-2 was fired, that was it; there was nothing you could do to change its trajectory and after *Brennschluss* (i.e., burnout), the missile was an unguided projectile. Grissom's Redstone, while controlled in a similar manner (i.e., internal and external vanes), was tracked, but not guided, real-time from the ground using the tracking radars, radar transponders, and onboard sensors. If the worst happened and the missile veered off course, what happened next was all up to the range safety officer.

Between September 6, 1944 and March 27, 1945, a total of 3,745 V-2's were launched by Germany against their enemies. Of these, 1,115 fell on England and 2,050 on targets on the continent. About ten percent of the rockets fired exploded in mid-air. It's estimated that the V-2 caused the deaths of 2,724 and injuries of 6,467 innocent civilians in England alone. However, the fuel feed and combustion chamber systems invented were instrumental in leading to the construction of many of the rockets used in both the United States and Soviet space programs.

Robert Goddard, after having success with his liquid fuel creations, died in 1945, largely unrecognized for his contributions to rocketry. Shortly before Goddard's death, his patent attorney attempted to collect royalties from the US Army for the post war use of the V-2's, since the German rocket had obviously infringed on Goddard's patents. It's not clear whether Goddard's estate ever made a dime off the V-2. After the end of the war, through Operation Paperclip, the left over V-2 components were shipped to the United States and assembled at White Sands, New Mexico for many months of tests and evaluations. With von Braun as their spokesman, more than 100 ex-Peenemünde engineers also emigrated to the United States with sufficient components to assemble about 100 rockets. It was at White Sands where on February 24, 1949, a modified V-2 rocket carrying a WAC-Corporal managed to reach an altitude of 244 miles, accomplishing the first true space shot. According to Willy Ley, ". . . at that altitude there are fewer air molecules in a cubic inch of air than in the best vacuum we can produce in our laboratories . . ." A man-made craft had finally probed empty space.

Besides creating the first medium range ballistic missile, the German military also significantly advanced aviation technology during the Second World War by manufacturing and flying the first operational jet and rocket powered aircraft (as did the Japanese, though on a lesser scale). While the Messerschmitt Me-262 jet aircraft did well against American bombers, it was placed in the field too late in the war to have a significant effect on aerial combat. In addition, Adolf Hitler made the ghastly mistake of trying to turn what was obviously the first jet fighter into a jet powered bomber, with disastrous results. The Me-163B Komet, a tiny rocket powered aircraft with a top speed of over 500 mph at sea level, gave US bomber crews quite a start when it was first sighted in the closing stages of the war. However, it had a landing approach speed of about 100 mph, which no doubt contributed to numerous accidents during landing, and it could only carry enough fuel to stay aloft for 20 minutes using minimum thrust; at full thrust, the time of powered flight dropped to a dismal 4 minutes!

The United States also started developing their first jet aircraft during the war when in 1942 Bell Aircraft test pilot Bob Stanley took the XP-59A Airacomet into the air following high speed taxi runs at Edwards Air Force Base. By 1943, the US had established an unofficial altitude record of 47,600 feet. In addition, in 1944, Lockheed's XP-80, later to become the F-80 Shooting Star, became the first American aircraft to exceed 500 mph in level flight. While none of these jet aircraft were used in combat during the Second World War, the F-80 later recorded the first jet aerial victory when it downed a MiG 15 in Korea.

In the late 1940's, as the supply of upgraded V-2's began to dwindle, various government and commercial organizations in the United States started developing a series of research rockets to continue and extend high altitude operations. The Aerobee, which was designed as a sounding rocket, was developed by the Applied Physics Laboratory at Johns Hopkins University; it could reach as high as 80 miles and was used as an upper atmosphere research tool for many years. The Naval Research Laboratory designed the Viking, manufactured by the Glenn L. Martin Company, which was later combined with the "Aerobee-Hi" in Project Vanguard, the United States' first attempt at launching an artificial satellite. Compared to the standard Aerobee, the "Hi" version had a slightly longer burning time (50 seconds as opposed to 34 seconds for the standard Aerobee) and a little more thrust (about 100 lbs.), allowing the rocket to reach an altitude of 122 miles with a 200 lb. payload, as opposed to 66 miles for the standard version.

Wernher von Braun and his team in Huntsville, Alabama, were also developing the Redstone, which used the same Rocketdyne rocket engine as the old Navaho booster. For guidance, it used an inertial navigation system with a gyroscopically stabilized platform, simple computers, and a pre-programmed flight trajectory. During the Redstone's maiden flight in August of 1953 at Cape Canaveral, the missile only made it 8,000 yards from the launch pad before crashing. Later on, the US Air Force began working on other new missile projects, among them the Atlas, Thor, and Titan, two of them to later play key roles in America's manned space flight program.

At the time, the idea of manned space flight was still something filling the pages of science fiction books as opposed to engineering design bureaus. However, by the late 1940's and early 1950's skilled test pilots were starting to fly very fast, and very high. First, Chuck Yeager, flying a Bell X-1 rocket powered test aircraft named "Glamorous Glennis," became the first man to travel faster than sound, in 1947, by exceeding Mach 1.06 (about 700 mph at 42,000 feet), crashing through what some called the sound barrier. Yeager also later accomplished the first (and only) ground takeoff of an experimental rocket plane by igniting the Bell X-1's rocket motors on the ground, only managing to reach an altitude of 23,000 feet before running out of fuel; a perfect example of why such test aircraft are normally ferried to altitude. Six years later, Scott Crossfield upped the ante by piloting the Douglas Skyrocket to a speed of 1,291 mph, becoming the first human to fly faster than twice the speed of sound. Finally, just after Crossfield's success, Yeager flew an improved Bell rocket plane in 1953, the X-1A, to over 1,650 mph at an altitude of 76,000 feet. It was also on this flight that the pilot encountered high-speed instability for the first time and narrowly escaped with his life as he fell more than 40,000 feet after losing control of his rocket plane. Test pilot Mel Apt wasn't as lucky in 1956 when his Bell X-2 tumbled out of control at over Mach 3 – he was killed.

While he wasn't a test pilot, a heroic figure during the early days at Edwards Air Force Base was Major John P. Stapp, who used himself as a test subject during numerous high speed runs on a rocket sled called the Sonic Wind. Prior to Stapp's experiments, conventional medical theories suggested that the human body could survive no more than 17 to 18 gravities of acceleration (or deceleration). In one 1951 test alone, the fearless Stapp was fired down a rocket sled track with 4,000 lbs. of thrust, crashing into a braking system (from 88.6 mph to a full stop in 18 short, incredible feet), causing him to experience about 48 gravities ("g's"); his body had absorbed an impact of over four tons. Stapp later explored even faster speeds and massive g loadings, some so severe that he was temporally blinded after the tests. He was finally ordered by his Air Force superiors to cease and desist the dangerous experiments, which probably would have eventually killed him.

Another Edwards program crucial to future manned space flight was the almost 200 test flights done in the late 1950's and mid-1960's with North American Aviation's X-15 rocket plane. Developed as a joint program governed by the Air Force, Navy, North American, and NACA, the sleek black rocket ship was designed to probe the region of flight from Mach 3.2 and beyond, or hypersonic flight. The basic flight characteristics of the aircraft were established by the NACA and the future Apollo spacecraft contractor was hired to build three X-15 research aircraft in September of 1955.

In many ways, the X-15 flights tested critical spacecraft hardware, isolated key design issues, and developed specific flight protocols that had a direct bearing on the success of later manned space missions. In particular, the X-15's hydrogen peroxide reaction control system was similar to what was later used on Mercury spacecraft, as was the chrome-nickel alloy, Inconel X (not much different than the René 41 nickel-steel alloy used on the exterior of Mercury spacecraft), used to withstand the temperature changes from launch through reentry. Other accomplishments were discoveries related to hypersonic boundary layer air flow (it was found to be turbulent and not laminar), turbulent flow heating rates (less than expected), and how space vehicle surface irregularities resulted in "hot spots" during reentry (this helped in the design of the Space Shuttle). In addition, the pilots flying the X-15s were outfitted with the first practical full pressure suit, ones not dissimilar from the suits worn by the Mercury astronauts. The rocket plane flights also established how to transition from aerodynamic controls to reaction controls and back again (as in Space Shuttle missions) and helped develop the energy management techniques needed for positioning winged space vehicles during landing approach.

Overall, the X-15 set a remarkable series of records during the years it flew at Edwards, such as the first flight past Mach 4 (Major Robert White, March 7, 1961), first time to an altitude over 200,000 feet (Major White again on October 11, 1961), as well the aircraft's highest mission, 354,200 feet above the Earth's surface (Joe Walker on August 22, 1963). With a length of about 50 feet, a launch weight of 31,275 lbs., fueled by anhydrous ammonia and liquid oxygen, a maximum thrust of 57,000 lbs. (with the XLR-99 rocket engine), and after being air dropped from an altitude of 45,000 feet, the basic X-15 managed to reach a maximum altitude of 58 nautical miles and speed of 4,520 mph (though not on the same flights). With a length of 83 feet, a launch weight of 65,940 lbs., fueled by alcohol and liquid oxygen, a maximum thrust of 78,000 lbs., and after being fired from sea level, Grissom's Mercury-Redstone combination reached a maximum altitude of 102.8 nautical miles and Earth fixed velocity of 4,512 mph, all during one mission.

Surprisingly, it was not at Edwards Air Force Base, but in Santa Monica, California, in 1946, where some serious thoughts were finally put to paper about the problems involved in sending a man into space. While the likes of Yeager and Crossfield were flying the wings off of everything they got their hands on, a small group of slide-rule carrying RAND (Research and Development) Corporation engineers were compiling a report called, "Preliminary Design of an Experimental World-Circling Spaceship:"

> *In this report, we have undertaken a conservative and realistic engineering appraisal of the possibilities of building a spaceship which will circle the Earth as a satellite. If a vehicle can be accelerated to a speed of about 17,000 mph and aimed properly, it will revolve on a great circle path above the Earth's atmosphere as a new satellite.*
>
> *— RAND Corporation, May 2, 1946*

"Satellites?" "Spaceships?" "Aimed properly?" That didn't sound much like a "real" aircraft like the Bell X-1A, and what was happening was the birth of a split of sorts with respect to the best road to space. Be that as it may, the

324 page RAND document addressed most of the major technical issues related to the construction of a space vehicle capable of placing a human into low Earth orbit. RAND envisioned a four-stage missile fueled by liquid hydrogen (identical to the fuel used on most of the Saturn V Moon rocket as well as the Space Shuttle), topped with a winged vehicle capable of reentering the Earth's atmosphere without burning up. In addition, the report discussed other potential problems such as the "method of guiding the vehicle on trajectory; the probability of striking a meteorite; controlling the temperature inside a spacecraft; pitch, yaw, and roll control in orbit; and the problem of descent and landing."

In the period between the end of the Second World War and the late 1950's, the approach to the technical challenge of space flight diverged into two camps: those who viewed space flight as a natural extension of high performance experimental aircraft and people who felt that the emerging rocket technology (and vehicles devoid of wings) was the way to go. In a way, both ideas had their strong points: If you had an aircraft, such as the X-15, that could already fly above much of the Earth's atmosphere (and keep the pilot alive while doing it), all you had to do was figure out a way to accelerate it past the critical 17,000 mph speed and presto, you were in orbit. When you wanted to land, assuming that you had figured out a way to slow down and descend back through the atmosphere without burning up, you could guide the aircraft down in a fairly conventional manner, much like the Space Shuttle lands today. To many people, it seemed to make sense: simply take a high-flying experimental aircraft and modify it to go all the way – into real space and orbit.

Unfortunately, the technical reality was different than what appeared to make sense. All of those wings, control surfaces, landing gear, and nose art end up weighing quite a bit – weight that needs fuel to get to altitude and accelerate to orbital velocity. If you got rid of all those wings and landing gear, you could use the weight for more fuel and perhaps other things. Also, even if you managed to build a winged craft that could get into orbit and keep the pilot alive, how were you going to get him down? If you didn't come up with some new materials, those sleek wings would end up being nothing but molten stubs of metal which would not be of much use to the pilot once he got back into the atmosphere, to say the least. And getting back through the Earth's thick atmosphere is a lot more difficult when you're going 17,000 mph, as opposed to 4,520 mph.

Everything changed, though, on October 4, 1957, when the Russians launched the first man-made satellite, Sputnik I, modifying the space-age playing field forever. Tass, the official Soviet news agency, had announced a couple of months before the Sputnik launch that they had tested a new ICBM having a thrust of over 400,000 lbs. That same missile was used in October to launch a 184 pound bundle of instruments called Sputnik, the words meaning "fellow traveler of the Earth." Only a month later, the Soviets sent the first living creature into space by launching Laika, the world's first space-traveling dog, into orbit on Sputnik II. Unfortunately, Laika's fame as an astronaut dog was short-lived, as the animal burned up along with Sputnik II during reentry. The spacecraft weighed 1,120 lbs., causing one U.S. general to comment, "We captured the wrong Germans."

We had not captured the wrong Germans. In reality, we had the wrong politicians in charge who, in their zeal to downplay the obvious competition with the Soviets, allowed the communist nation to make space history. While we were configuring untried rocket hardware like the Vanguard, field proven rockets like the Redstone were shoved out of the picture. It all came to a head when on December 6, 1957 and in front of a national television audience, the Vanguard's first stage exploded in a massive conflagration spreading rocket fuel and bits of missile all over the launch pad. Fortunately, it wasn't long before saner heads asked Wernher von Braun and the Redstone Arsenal to dust off "Project Orbiter," a scheme involving the use of a Redstone with clustered upper stages to shoot a small instrument package into orbit. Incredibly, only 84 days after being officially "turned on," a Jupiter C rocket (actually an elongated Redstone) lifted the United States' first satellite, Explorer I, into orbit on January 31, 1958. The space race was in high gear.

In 1958, the same year NASA was officially activated, the fastest a human being had flown was 2,094 mph and the highest, 126,200 feet (a little over 20 nautical miles). That was a long ways from 17,000 mph and 102 nautical miles up. However, serious efforts were in progress to change all that, whether it be by using winged aircraft or what President Eisenhower called "basketballs in space." Even with that, an important question is, why bother? Was there any pressing reason for going through the trouble to explore space? Why spend the money and risk lives to do it at all?

There were some legitimate reasons to develop a space capability. First of all, it's a lot easier to exploit space as a potential battlefield if you have the ability to get there to begin with, and in spite of NASA's public persona as a "civilian" organization, much of the technical and operational support for NASA's civilian manned space program initially came from, guess what? The military. The men who would eventually become Mercury astronauts, were all military pilots. The missiles used in early manned space missions were all converted tactical and strategic missiles. The early launch pads were actually part of the Cape Canaveral Air Force station. At least part of the Mercury tracking network was supplied by the Air Force's Eastern Missile Test Range. And all of the recovery ships were supplied by the US Navy. As a result, it was a bit of a stretch to call NASA's space activities strictly "civilian efforts," and to this day, the military continues to play a significant role in the agency's program.

Given the political "black eye" President Eisenhower received after the Soviets launched Sputnik I, his administration was definitely motivated to strike back, from the technological standpoint. What would it look like if a communist country was able to demonstrate technological superiority over the United States? It wouldn't say much for democracy, that's for sure. Consequently, the United States was destined, if not obligated, to respond to any and all communist demonstrations of technical prowess. The media made it even worse by claiming that a "missile gap" existed between the United States and the Soviet Union (implying that the Russians had the big boosters and we did not), an issue that Presidential hopeful John F. Kennedy used to great benefit during his political campaign. How could we claim that democracy was a superior form of government when a communist country was clearly ahead of us in the most sophisticated form of technology on the planet? We couldn't.

Finally, in the late 1950's and early 1960's, space was the "new ocean." We, as human beings, had already conquered the atmosphere using balloons, propeller driven aircraft, jets, and rocket planes. We had explored the oceans on the surface with ships and underneath with submarines (though even today, most of the ocean remains unexplored). We had already mapped most of the globe. Space was the only place left to go. It was natural that, as a people, Americans wanted to continue exploring the limits of our environment. From the political standpoint and taking the Cold War into account, it was an easy sell to Congress and the American taxpayers.

At the time, there was more than one option for getting a man into space. First, there was the Air Force's "Man in Space Soonest" (MISS) proposal, under development by their Air Research and Development Command. They had

High School: Virgil I. Grissom at high school age. As a youth, the future astronaut was fascinated by the sounds of power and spent many hours building model airplanes. While he was only an average student, he had an IQ of 145, which put him in the gifted range. Photo courtesy of Wilma Beavers.

In the Air Force: Grissom, along with two friends, was inducted into the Air Force just after high school. Even though he wanted to fly, he was stuck being a clerk typist at Brooks Field, in San Antonio, Texas. It would not be long before he got what he wanted, and more. Photo courtesy of Wilma Beavers.

Solo Flight: The future test pilot with his propeller-driven trainer on the day he flew alone for the first time. The notation on the back of this photograph states, "This was taken the day I soled Jan. 16, 1950 see I have the zipper of my flying suit down, you can't have it down until you solo." Photo courtesy of Wilma Beavers.

Learning to Fly: Grissom at the controls of what is probably an T-6 "Texan" during primary flight training. This was after he graduated from Purdue University with a degree in mechanical engineering. Photo courtesy of Wilma Beavers.

Jet Pilot: The new fighter pilot standing in front of a Lockheed F-80 "Shooting Star" jet fighter. This picture was probably taken when Grissom was making the transition from propeller to jet powered aircraft. Photo courtesy of Wilma Beavers.

On the Way to Korea: Grissom (far left), with fellow F-86 fighter pilots at Presque Isle, Maine, just before being shipped over to Korea. This particular jet fighter was called the "Hadacol Special" and belonged to fighter pilot William Wood Senior (second from right). Grissom's jet in Korea was named "Scotty," after his son, Scott Grissom. Photo courtesy of William Wood, Jr.

In Country: Air Force fighter pilot Virgil I. "Gus" Grissom departing the famous "Swig Alley" Officers Club at Kimpo Air Base in Korea, home of the 4th Fighter Interceptor Wing. Grissom was denied many opportunities to shoot down a Mig, mostly because he was required to fly wing for the squadron commander. By the time he left Korea, Grissom had flown 100 combat missions. Photo courtesy of Wilma Beavers.

F-104 Driver: One of Grissom's favorite aircraft was Lockheed's F-104 Starfighter, which is still in use today by some foreign air forces. Judging from Grissom's rank of Captain, this photograph was probably taken in the late 1950's or early 1960's. The F-104, which was developed by famed designer Kelly Johnson, was the first operational interceptor capable of sustained speeds above Mach 2 and was also known by the nickname "missile with a man in it." Photo courtesy of Wilma Beavers.

Family Home: Grissom's boyhood home in Mitchell, Indiana, as it exists today. C. Newport Photo

Grissom the Astronaut: Virgil I. Grissom's official NASA portrait. Even though Grissom had reservations about joining Project Mercury at first, the opportunity to fly higher and faster than anyone else was too great. NASA Photo.

Fellow Mercury Astronauts: Grissom (left) in 1959 with fellow Mercury astronauts Scott Carpenter, Deke Slayton, and L. Gordon Cooper (front) and Alan Shepard, Walter M. Schirra, and John Glenn (back). NASA Photo.

Dr. Robert Goddard: Professor Robert Goddard of Worcester, Massachusetts, with the first successful liquid fuel rocket. The device traveled a distance of 184 feet in 2.5 seconds at about 60 miles per hour. NASA Photo.

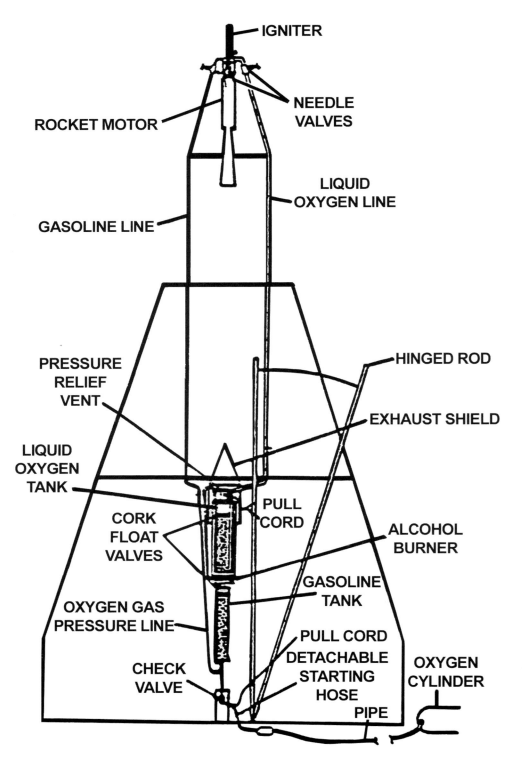

Liquid Fuel Rocket: Even though Goddard's creation was very simple, it did prove that the more powerful liquid fuels, when used in combination with an oxidizer, could be made to work. NASA Photo.

Up in Smoke: America's hopes for an artificial satellite get turned into ashes in front of a national television audience on December 6, 1957. It wasn't long before "old reliable," the Redstone rocket, was dusted off and modified into a successful satellite carrier. NASA Photo.

A Hard Won Success: Only 84 days after being officially turned on by the government, Project Orbiter's Jupiter C (an elongated Redstone) successfully inserted the Explorer I satellite into orbit on January 31, 1958. NASA Photo.

Ground Test: A JF-104A Lockheed Starfighter undergoes a ground test of its Reaction Control System (RCS) at the Dryden Flight Research Center sometime in 1961. It was this sort of technology development that supported manned spacecraft design. NASA Photo.

Dr. Robert R. Gilruth: As director of NASA's Space Task Group at Langley Research Center in Virginia, Gilruth was instrumental in dealing with the complex technical and management issues that made Project Mercury a success. His keen knowledge management style allowed his engineers and scientists to isolate and overcome the significant technical hurdles that made manned space flight a reality. NASA Photo.

Rocket Scientist: Dr. Maxime Faget (foreground), the designer of the Mercury spacecraft, works with Apollo astronaut Frank Borman in a mockup of the Apollo Command Module in 1967 during the Apollo 204 Review Board. The Board investigated the fire that took the lives of astronauts Grissom, White, and Chaffee earlier that year. NASA Photo.

Mercury in Orbit: This artist's rendering shows what a Mercury spacecraft might look like in low Earth orbit as it circles the globe. Visible on the exterior of the craft's heat shield are the three retrograde and posigrade (the smaller nozzles in between the retros) solid fuel rocket motors. Boeing Illustration

Master of Reentry: Senior aeronautical engineer Harry Julian Allen with one of his test models, probably at the Ames Research Center. Allen's common sense approach to the physics of reentry vehicles was directly responsible for the development of the Mercury spacecraft's blunt heat shield design. NASA Photo.

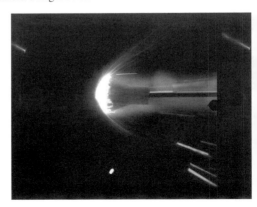

Shock Wave: This scale model of a Mercury spacecraft clearly shows how a shock wave sitting just in front of the heat shield could help dissipate the heat generated during reentry. It was Julian Allen's concept that a blunt aerodynamic object would be more efficient from the thermal standpoint than a more conventional shape. NASA Photo.

Little Joe in Flight: One of several Little Joe rockets tears away from the launch pad at NASA's Wallops Island facility early in Project Mercury. Using these test vehicles, which were configured from a clustered set of Pollus and Recruit rockets, offered a far more cost effective way to validate the Mercury parachute and escape rocket systems. NASA Photo.

Not as Planned: The escape tower from the Mercury-Redstone 1 dug itself a nice grave after the Redstone shut down prematurely on November 21, 1960. NASA Photo.

Mangled Spike: The aerodynamic spike from the MR-1 escape rocket has obviously seen better days. The rocket landed about 1,200 feet from the launch pad during the 1960 test. NASA Photo.

Escape Systems Sequence: This series of images shows the nominal chain of events following the ignition of a Mercury escape rocket, in this case, from the pad at Wallops Island. After the rocket's fuel burns out, the tower is jettisoned, followed by parachute deployment. Fortunately, it was never necessary to use the escape system with a live astronaut. NASA Photo.

Mercury Parachute Deployment: This sequence shows the deployment of a capsule's ring-sail parachute. Initially, the drogue chute is deployed, following by the reefed main chute, which opens fully a few seconds later. NASA Photo.

Wind Tunnel Tests: A full scale spacecraft mockup sits on a test stand at the Langley Research Center in Hampton, Virginia. One of the best design features of the Mercury spacecraft was that it was passively stable, from the aerodynamic standpoint. NASA Photo.

Mercury Spacecraft Concept: The Mercury spacecraft, when coupled with the tubular escape tower, looked anything but aerodynamic. The word agricultural comes to mind. Boeing Illustration.

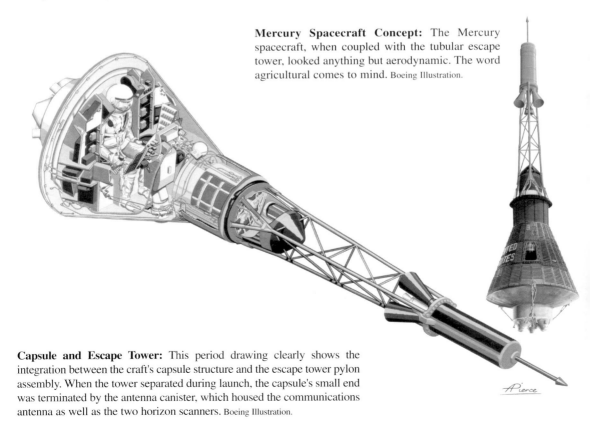

Capsule and Escape Tower: This period drawing clearly shows the integration between the craft's capsule structure and the escape tower pylon assembly. When the tower separated during launch, the capsule's small end was terminated by the antenna canister, which housed the communications antenna as well as the two horizon scanners. Boeing Illustration.

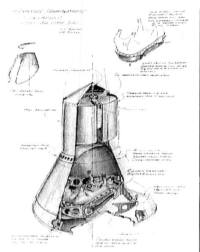

Pig Drops: It was essential to establish that a human could withstand the impact of a Mercury capsule on land, which was the logic behind the famous "pig drop" experiments depicted here by Caldwell C. Johnson. NASA Drawing.

Grissom with MASTIF Trainer: The MASTIF (Multiple Axis Space Test Inertia Facility) was located at the Lewis Research Center's altitude wind tunnel. Grissom's first meeting with the MASTIF was in February of 1960. It's not known who won. NASA Photo.

MASTIF at Work: Both Grissom and Shepard endured a few runs in the MASTIF and it was eventually verified that 30 revolutions per minute in three axes was the limit of human tolerance. During Shepard's first run he quickly turned green and hit and hit the red "chicken switch." The contraption was definitely not for the squeamish. NASA Photo.

Boilerplates: NASA technicians at the Langley Research Center put the finishing touches on a boilerplate Mercury spacecraft. While the interior of the craft held only instrumentation, these vehicles, when coupled with the Little Joe booster, offered an inexpensive way to evaluate specific design parameters. NASA Photo.

Drop Tests: The boilerplate spacecraft were also useful for establishing the stability of the craft during the floatation phase. While the boilerplates were only shells, they had most of the weight and balance characteristics of a full up space capsule. NASA Photo.

Retrograde Package: A technician works on a Mercury spacecraft heat shield already fitted with the solid fuel retro and posigrade rockets. The electrical cabling was used for the ignition system and thermal blankets. NASA Photo.

McDonnell Aircraft White Room: Several early Mercury spacecraft are shown under construction at McDonnell's St. Louis, Missouri facility in this April 1960 photograph. What appears to be an escape tower pylon is situation in between capsules 2 and 5. Boeing Photo.

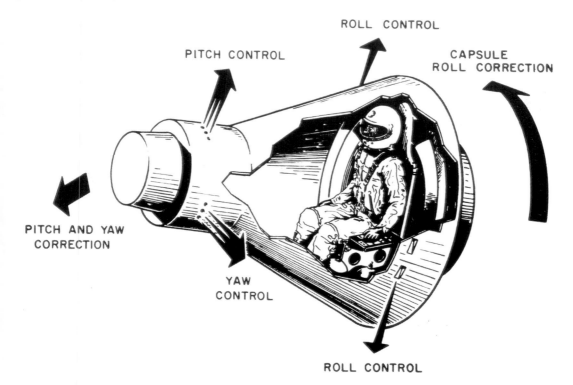

Capsule Control: This early NASA drawing shows the various control axis as well as the placement of the Reaction Control System thrusters on a Mercury spacecraft. NASA Illustration.

Impact Landing System: NASA personnel collapse and install a spacecraft landing bag, used to attenuate the capsule's impact with the sea or land. The original design of the landing bag was devoid of external straps and any internal tensioning cables. However, field tests during actual flights illustrated this weakness in the design. NASA Photo.

Centrifuge Training: Astronaut Grissom with Dr. William Augerson just before entering the US Navy's centrifuge at Johnsville, Pennsylvania. NASA Photo.

Under Construction: Three Mercury capsules sit in special frames during fabrication in 1961 at McDonnell's facility. The vehicle's titanium framework is clearly visible as are the cylindrical parachute compartments. Boeing Photo.

In the Gondola: Grissom is strapped in tight during one the centrifuge runs in Johnsville prior to his Liberty Bell 7 flight. The Navy Medical Acceleration Laboratory was capable of accelerations as great as 18 g's. NASA Photo.

On His Way: America's first man in space, Alan Shepard, lifts off on Mercury-Redstone Flight No. 3 on May 5, 1961. The mission was an unqualified success and paved the way for Grissom's follow up attempt. There was actually a third suborbital flight scheduled, which was canceled to make way for Glenn's orbital mission. Boeing Photo.

Water Egress: Astronaut Wally Schirra tries his hand at exiting a Mercury boilerplate while floating in the water. Scott Carpenter was the only astronaut to actually exit a spacecraft in this manner; all the rest of them used the side hatch. NASA Photo.

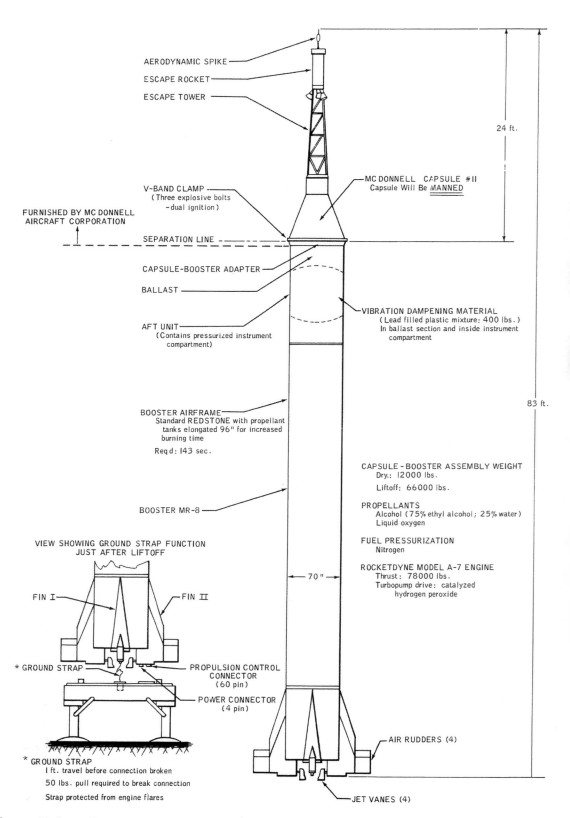

Mercury Redstone Booster: NASA diagram of the modified Redstone tactical missile used on all suborbital flights. NASA Drawing.

Mercury Control Center (MCC): Before the Johnson Space Center was built in Houston, Texas, all NASA manned missions were run out of the MCC, a cinder block building not far from Grissom's launch pad. The plot board shows the real-time location of a Mercury capsule in orbit and it was here that the term "Cap Com" originated, which is still used and stands for Capsule Communicator. NASA Photo.

Periscope Inspection: Virgil Grissom checks out the optical periscope on a Mercury spacecraft, probably in the white room at the top of launch pad 5. NASA Photo.

Water Training: The astronaut in a one man mylar life raft during survival training sometime before his Mercury mission. Little did he know how important his water survival skills would be during his flight. NASA Photo.

Simulation: Astronaut Grissom is helped out of Liberty Bell 7 by McDonnell Aircraft personnel following a flight simulation on July 17, 1961. NASA Photo.

Arrival by Air: The Redstone booster scheduled for the Mercury-Redstone No. 4 mission gets unloaded from an Air Force transport aircraft in June of 1961. It would not be long before similar rockets used for America's space program would need whole barges to be transported to the Cape. NASA Photo.

asked for and received numerous proposals from government contractors suggesting various ways to attempt manned orbital flight (the key word here is, attempt); they ran the gamut: North American, the builders of the famed X-15 rocket plane, not surprisingly, offered a winged craft, called the Dynasoar Approach." AVCO Manufacturing Corporation proposed a spherical craft fitted with a steel mesh parachute of sorts, making the whole thing look much like a shuttlecock. Both Bell and Republic Aircraft suggested more winged spacecraft, with Bell saying that the use of a "wingless" craft "would be only a stunt." Lockheed, Martin, McDonnell, Goodyear, and Convair, all proposed some sort of elongated or spherical craft, suggesting that they be launched using Atlas, Titan, or Polaris missiles (or some combination thereof).

The Army had a program in the works initially called "Man Very High," later renamed "Project Adam." This was led by Wernher von Braun and his associates at the Redstone Arsenal and suggested using a modified Redstone booster in combination with a space "capsule" along a very steep ballistic trajectory. Unlike his earlier work in Germany during the war, von Braun didn't have to worry about a Heinrich Himmler throwing him in jail for thinking about space travel. According to their calculations, the capsule might reach an altitude of about 150 miles before landing about the same distance from Cape Canaveral along a suborbital flight path. Incredibly, the Army tried justifying the plan as a possibility for improving methods of transporting troops! If you can imagine launching Army draftees into space it's easy to see that the idea was not that feasible, at least from the standpoint of mobilizing government personnel. Not surprisingly, Hugh Dryden told the House Space Subcommittee that "tossing a man up in the air and letting him come back. . . is about the same technical value as the circus stunt of shooting a lady from a cannon." However, the first two Mercury shots ended up being just that: suborbital flights from Cape Canaveral with the next stop the Atlantic Ocean. Someone must have changed their mind along the way.

A third possibility was the Navy Bureau of Aeronautics' "Manned Earth Reconnaissance" or MER I project. They envisioned an orbital mission using a cylindrical spacecraft having two spherical ends. What was unique about the approach was that after being fired into orbit, the vehicle's ends expanded to make a delta-winged inflatable glider (a water landing was planned).

It wasn't long before all of the above ideas ended up being just that, ideas, as President Eisenhower decided to place the responsibility for manned space flight into the hands of a new government agency, the National Aeronautics and Space Administration. This historic decision, much to NASA's satisfaction, also resulted in the transfer of $53.8 million of Air Force monies to the fledgling agency, along with other funds from the Department of Defense and the Air Research and Development Command. Even so, the Air Force elected to continue working on their Dyna-Soar project.

Fortunately for NASA, they had some smart individuals who would ensure that the new manned space program was headed in the right direction: Among them, Dr. Robert Gilruth, who became the Director of the Space Task Group (originally called the Manned Ballistic Satellite Task Group) at the Langley Research Center and Dr. Maxime A. Faget, head of the Langley Research Center's Performance Aerodynamics Branch in the Pilotless Aircraft Research Division (PARD). Gilruth, the former chief of PARD, came to Langley with a master's degree in aeronautical engineering from the University of Minnesota and was a specialist in flight research and guided missile development. He was well known in the industry as not only a talented scientist, but also an effective manager of complex, large-scale engineering programs.

Given Gilruth's working relationship with Faget, it came as no surprise that the ex-submariner would be asked to figure out the complexities of developing America's first manned space vehicle. Faget, born in British Honduras and after graduating with a degree in Mechanical Engineering from Louisiana State University, originally wound up underwater as a junior officer on a US Navy submarine during the War in the Pacific. However, once he got his feet back on dry land in Langley, Virginia, he was instrumental in developing the basic concept of the Mercury spacecraft and pioneered the use of real-time telemetry during flight tests. Others who shared pivotal roles in the development of the Mercury spacecraft were William Blanchard, Robert Chilton, Caldwell C. Johnson, Alan Kehlet, Jerome Hammack, and Andre Meyer. In particular, Caldwell C. Johnson, a self made engineer, worked with Faget to refine the Mercury capsule design during its development at Langley (he also created some accurate and fascinating engineering sketches of early space hardware).

Even before NASA formally sent out a formal Request For Proposals (RFP), James S. McDonnell, Jr., the founder of a growing aerospace contractor bearing his name, was hard at work on a path parallel to Faget's. What's interesting is that in 1957 during a speech at an engineering school's commencement ceremony, McDonnell predicted that the first manned space satellite, weighing four tons and costing a billion dollars, would not be accomplished until sometime between the years 1990 and 2005 AD! Fortunately, for McDonnell Aircraft's financial sake, he was off by over 40 years. In spite of McDonnell's conservative estimates for the timetable of manned space flight, he had over 70 men in St. Louis working hard on a preliminary design for a manned satellite. Among these men were John F. Yardley and Lawrence Weeks, both to figure prominently in the creation of a 427 page prospectus as well as the actual construction of the first American manned spacecraft. Their early efforts proved to be worth the time and expense.

In late 1958, technically, NASA's manned space effort was simply called the "Manned Satellite Project." Langley suggested to headquarters three possibilities for simple insignias that could be used in conjunction with the effort: symbolic images such as Phaeton and Apollo, the Great Seal of the United States, or a map of the globe. The Space Task Group, the component of NASA responsible for future manned space missions, thought "Project Astronaut" was appropriate, considering the manned aspect of the effort. Ultimately, Abe Silverstein, appointed as the Director of Space Flight Development, won out in the end, advocating the Olympian messenger Mercury, a "Roman deity and messenger of the gods, god of commerce, dexterity and eloquence." Project Mercury it was.

It was not until September of 1958 that the new NASA organization began to formally establish what they were going to do and how they were going to do it. This was done by the Joint Manned Satellite Panel, which was made up of personnel from both the old NACA and ARPA (Advanced Research Projects Agency). The joint NACA-ARPA panel was established by T. Keith Glennan and Roy W. Johnson and was composed of six NASA and two ARPA members. Their outline report, developed with the assistance of thousands of scientists and engineers, was entitled "Objectives and Basic Plan for the Manned Satellite Project." This finally dictated what NASA had in mind to get an American into space:

I. Objectives: The objectives of the project are to achieve at the earliest practicable date orbital flight and successful recovery of a manned satellite, and to investigate the capabilities of man in this environment.

II. Mission: To accomplish these objectives, the most reliable available boost system will be used. A nearly circular orbit will be established at an altitude sufficiently high to permit a 24-hour satellite lifetime; however, the number of orbital cycles is arbitrary. Descent from orbit will be initiated by the application of retro-thrust. Parachutes will be deployed after the vehicle has been slowed down by aerodynamic drag, and recovery on land or water will be possible.

III. Configuration (Vehicle): The vehicle will be a ballistic capsule with high aerodynamic drag. It should be statically stable over the mach number range corresponding to flight within the atmosphere. Structurally, the capsule will be designed to withstand any combination of acceleration, heat loads, and aerodynamic forces that might occur during boost and reentry of successful or aborted missions.

— NASA Joint Manned Satellite Panel, October 1958

What they did was lay the groundwork for America's first effort to send a man into space and return him safely to Earth, an effort that would eventually involve thousands of technicians, engineers, and scientists, numerous small businesses and corporations, three branches of the military, ships, aircraft, and even several foreign countries. In addition, in a few paragraphs of text, NASA had established the general configuration of the first United States manned space vehicle and it would eventually cost $400,658,000 to achieve these goals. Starting in October of 1958, NASA's personnel would explode from the only 45 people initially assigned to the project, to 3,345 scientists and engineers; the Space Task Group would quadruple in size, and seven unknown military pilots would become household names.

One problem NASA had was where to find the men needed to pilot their as yet undesigned craft into an environment never before visited by humans. Their name, "Astronauts," was derived from the pioneers of ballooning, who were called "Aeronauts." This was identical to the legendary Greek "Argonauts," who sailed into uncharted oceans in search of the Golden Fleece. In more realistic terms and as noted in their civil service job specification, NASA and the Space Task Group described their required duties as:

> He will contribute by monitoring the cabin environment and by making necessary adjustments. . . have continuous displays of his position and attitude. . . will have the capability of operating the reaction controls, and of initiating the descent from orbit. . . will contribute to the operation of the communications system. . . and make research observations that cannot be made by instruments. . .
>
> — Specifications for Research Astronaut-Candidate, NASA Space Task Group, December 22, 1958

Most importantly, NASA's prospective spacemen had to have demonstrated, ". . . a willingness to accept hazards comparable to those encountered in modern research airplane flight and (have the) . . . capacity to tolerate rigorous and severe environmental conditions and ability to react adequately under conditions of stress or emergency." Overall, NASA was not certain exactly who they were looking for when they published "NASA Project A, announcement No. 1," in 1958 for the position of "Research Astronaut-Candidate," which offered a starting salary of $8,330 to $12,770, depending upon the applicant's qualifications. Besides the obvious choice of professional or military test pilots, they were also considering individuals with experience in commercial aircraft, ballooning, submarines, or even medical doctors and others who simply had three years of work experience in any of the physical, mathematical, biological, or psychological sciences. To help rule out "nut cases" NASA also required that the prospective candidate have the sponsorship of a "responsible organization."

However, in spite of NASA's public guise as a civilian agency, President Eisenhower later ordered them to select their "Research Astronaut Candidates" from the existing group of active duty military test pilots. The submariners, balloonists, and scientists were out and fighter pilots were in. To find the men having the appropriate background, NASA submitted a simple list of criteria to the Pentagon who later replied that 100 men on active duty appeared qualified. Even while some of the applicants felt that there would be little actual "piloting" during a Mercury mission, most experienced test pilots know that no craft is foolproof and would have to be controlled real time during at least some portions of the flight.

By December 29, 1958, NASA had completed reviewing the 11 responses to their RFP for the Manned Satellite Project. Max Faget, who developed the overall concept for the manned ballistic capsule, was very specific in the wording of the document informing prospective bidders to, ". . .incorporate the retrorocket principle, the non-lifting principle, and the non-ablating heat sink principle." These criteria essentially restricted proposed designs to a wingless space vehicle using rockets to deorbit, and some sort of heat shield to protect the astronaut from heat during reentry. However, in what almost appeared to be a contradiction, Faget also "invited the bidders to submit alternate capsule and configuration designs." As a result, four of the companies were disqualified on technical grounds. In the end, the competition came down to two companies:

McDonnell Aircraft Corporation and Grumman Aircraft Engineering Corporation. McDonnell eventually got the job, mostly because NASA thought Grumman was too busy with Navy work to handle the additional responsibility. In addition, NASA's Manned Satellite had been granted "DX" status (i.e., highest industrial procurement priority on the Department of Defense Master Urgency List). As a result, as far as NASA was concerned that a, ". . .serious disruption in scheduling Navy work might occur if the higher priority capsule project were awarded to Grumman."

In short order, McDonnell Aircraft engineers were expanding on Faget's 50-page "Specifications for a Manned Satellite Capsule," where the Space Task Group engineer outlined his concept for using a non-lifting craft to orbit a man above the Earth in space. McDonnell was also contractually obligated to construct a dummy Mercury capsule by the spring of 1959, even though the contractor and NASA engineers did not officially start working together until January of that year.

Besides Virgil I. "Gus" Grissom, six other men were also selected by NASA to take the first American rides into space: M. Scott Carpenter, L. Gordon Cooper, John H. Glenn, Walter M. Schirra, Alan B. Shepard, and Donald "Deke" Slayton. All of these men found out about NASA's new program in various ways: Carpenter was serving as an intelligence officer on an aircraft carrier when he got a message saying that he would, ". . .soon receive orders to OP-05 in Washington in connection with a special project." Carpenter was unhappy about being assigned to sea duty, especially work that chained him to a desk. When he finally got on dry land, he was shocked to read a Time magazine describing how, "110 pilots with certain qualifications were being ordered to the Pentagon to see if they were qualified for manned space flights." Carpenter later recalled, "Good Lord! That couldn't possibly be me!" It was. After sitting through NASA's briefing in Washington, Carpenter was eventually told that the new agency wanted him on board for Project Mercury. That fact, however, did little to impress the Captain of the carrier who knew nothing of Project Mercury or space travel; until he received official orders to the contrary, he expected Carpenter to go back to sea as the ship was ready to sail. Carpenter's boss, however, was understanding in that he might let the astronaut go as long as he was back in time for maneuvers; he wasn't.

L. Gordon Cooper, assigned to Edwards Air Force Base at the test pilot school, remembers reading one day that the McDonnell Aircraft Company in St.

Louis had been awarded a contract to build a space capsule. As Cooper remembered, "This really interested me. Flying has always been one of the most fascinating and satisfying things in the world to me. . . There appeared to be a possibility that we might fly in space. . . I could almost taste it." A few days after Cooper read the article about the space capsule, he was called to Washington for a classified briefing on Project Mercury. After the briefing and when asked his reaction to what he had seen and heard, Cooper said he was definitely sold on the program.

The politically astute Glenn, a Major and pilot in the Marine Corps, was working at the Patuxent Naval Air Station and in Washington, D.C., when he heard the word on the street about Project Mercury. As the astronaut remembered, ". . . our office was asked to furnish a test pilot to visit the NASA laboratory at Langley Air Force Base in Virginia and make some runs on one of the space flight simulators as part of a NASA investigation of various reentry shapes." Glenn managed to get the job and started thinking that maybe he might want to try and get in on the new government program. After most of the prospective military pilots had been screened out, Glenn's name was still in the running; he completed the physical and psychological tests and even though he was older than the other candidates, was asked to join the program. Glenn accepted immediately.

In 1958, Schirra was assigned to the Patuxent Naval Air Station as a test pilot, checking out the F4H Phantom aircraft, the first one off the line. However, in early 1959, like many of his peers, he received orders to report to Washington for a new and unknown program; "The suspense lasted until the day I reported to the Pentagon along with thirty or forty fellow officers. We learned that we were candidates for a program of the National Aeronautics and Space Administration to send a man into space," the pilot later remembered. Schirra, in particular, was dubious about leaving the Navy for the NASA project; "I wanted to be cycled back to the fleet with the F4H, get credit or take blame for its performance, and put it through its paces as a tactical fighter." However, in spite of his misgivings, Schirra could not turn down a chance to be first at something, whether it be in space or landing on the Moon. He was in.

Alan Shepard was happy with his future in the Navy when he first heard of Project Mercury. The Navy pilot was a staff officer at Atlantic Fleet Headquarters in Norfolk, Virginia and responsible for aircraft readiness. He thought his chances of becoming the skipper of a carrier squadron were excellent. "When Project Mercury came along... I was fairly well satisfied with my prospects, though I was by no means complacent. Like some of the others, I had quite a debate with myself about what to do," Shepard recalled. Even though Shepard consulted with his wife, Louise, she replied, "Why are you asking me? You know you'll do it anyway." Shepard had read about Project Mercury long before it was officially announced, and after consulting with his boss and being offered the job, he couldn't turn it down. "It was a Thursday morning when the word came... It was Mr. Donlan at NASA... he asked me if I was still interested in being an astronaut. I told him that I was... the office was empty – I let out a loud whoop."

By Slayton's accounting, he "had the best job in the Air Force when Mercury came along." Slayton had been working as a test pilot at Edwards Air Force Base for about four years; "This was the one place in the Air Force where I could utilize both my engineering background and my piloting experience, and put it all together in what seemed to me the ultimate . . . so I was not too eager to leave there." However, the test pilot soon realized that his "fun" was going to end soon, whether he liked it or not, as the Air Force had just created a new regulation limiting him to five years in any one spot. Even so, when Slayton was called to Washington, he was unsure about NASA's new program and whether or not they even needed a trained pilot. After all, simply sticking some guy on the end of a missile and launching him into space didn't sound much like real flying. After hearing NASA's pitch, though, Slayton came away convinced that they couldn't do what they wanted to do *without* a experienced test pilot. After Slayton was asked to join Project Mercury, he realized that it was "a test pilot's dream," and was happy to be in on the ground floor of the new project.

NASA was faced with a daunting task. Already, the United States had been slammed by the Russians several times after the communist country launched the first artificial satellite and living creature into orbit, and even fired a 3,245 lb. spacecraft towards the Moon. The craft, called Mechta or Lunik I, was the first man-made object ever to leave the Earth's gravitational field. Even though the Russians missed the Moon, it was clear that Soviet Premier Nikita Khrushchev was determined to use the new "high ground" of space as a way to demonstrate communist technical superiority. It would not be long before they had a man in orbit.

Even though Faget and the rest of Langley's Space Task Group had already figured out the basic configuration for their "Mercury Spacecraft," now they had to actually build it. It was one thing to have pretty pictures of theoretical engineering concepts and quite another to actually construct a craft capable of carrying a man into space and returning him back to Earth, preferably, alive. The basic theories and general concepts had to now be defined to a level of detail where a McDonnell Aircraft engineer could create construction drawings that could be used by technicians to manufacture actual flight hardware: components that a person could hold and touch, as opposed to high-level scientific concepts that could only be explained by word or image.

In general, the Space Task Group had to overcome several technical and operational hurdles. First, they had to develop a space vehicle light and strong enough to be accelerated to a speed of about 17,000 miles per hour above the atmosphere by an Atlas booster. In addition, that same craft had to sustain the life of a living, breathing man within very specific limits of temperature, pressure, and gravities; incorporate an as yet undesigned subsystem that would allow the pilot to orient the craft during flight (especially during reentry); provide some method of communication and tracking; and allow ground personnel the opportunity to know what was happening with both the space vehicle and the test pilot.

They also had to have at their disposal numerous resources to make it possible to track the spacecraft during the final stages of flight, such as aircraft, ships, and helicopters, and after landing, collect up the astronaut and his spaceship and transport him back to the United States. However, first things first: they had to design, build, test, and fly a Mercury spacecraft to see if any of it was actually possible. They had to build America's first spaceship.

Chapter 3 – The Mercury Spacecraft
Building and Testing America's First Spaceship

A good scientist is a person with original ideas. A good engineer is a person who makes a design that works with as few original ideas as possible.
— Freeman Dyson

The basic shape and design of America's first manned spacecraft did not exactly happen by accident. First of all, there were the likes of Dr. Robert Gilruth, Dr. Max Faget, and Caldwell C. Johnson driving the engineering trade studies with their individual ideas about the most efficient way to get an American into orbit. Surprisingly, Johnson's earlier rough engineering drawings showing an American spacecraft in various stages of flight were not too different from what NASA eventually built. Overall, if NASA was interested in using existing rockets as launch vehicles, then they were limited to certain criteria that ultimately determined the size and weight of a Mercury spacecraft: it had to fit on top of an Atlas booster and be light enough to be accelerated to 17,000 mph. If the spacecraft did not fit within those guidelines, it was not going to work; at least not with America's existing stable of guided missiles.

When McDonnell Aircraft won the NASA Project Mercury contract, only a few design parameters had been established by Dr. Faget: the spacecraft would have a non-lifting or blunt shaped exterior; it would be fitted with a "heat sink" to protect the structure and astronaut during reentry; and the craft would be de-orbited using retrorockets (i.e., small solid fuel rockets oriented towards the direction of flight), which, when fired, slowed the capsule down sufficiently that it would drop back into the atmosphere.

The first order of business for McDonnell and NASA's Space Task Group (STG) was to finalize the engineering concepts for the spacecraft's thermodynamic protection, and escape and parachute systems. Even though NASA's RFP specified the use of a "heat sink" (i.e., a heat radiating metal as opposed to an ablative material) for the spacecraft's heat protection, some at NASA were not certain it was the way to go for the new manned satellite. As a result, the heat sink approach formed one half of a parallel research program designed to establish the most efficient method of heat protection during reentry. Unfortunately, NASA engineers predicted that the expensive metal component would get pretty damn hot during reentry, hot enough to potentially cook the astronaut inside before landing. As STG member George M. Low reported at a Langley meeting in January of 1959:

> *It is anticipated that flights with both types of heat protection will be made. .. In case of a recovery on land, the capsule with a beryllium heat sink will require cooling; this is accomplished by circulating air either between the heat sink and the pressure vessel, or by ventilating the pressure vessel after impact.*
> — George M. Low, NASA Space Task Group, January, 1959

There were even concerns that the blistering hot heat sink might start a brush fire if the capsule landed in a field or forest (it was predicted that the heat sink might encounter temperatures as high as 3,500 degrees F during reentry). As a result, experiments were done to explore the feasibility of jettisoning the shield before landing. Unfortunately, during an early test, the detachable heat shield went into a classic falling leaf pattern and promptly crashed back into the descending capsule. Clearly, that approach was not going to work. There were also problems with even finding a sufficient supply of the exotic metal to manufacture the heat sinks. In fact, in 1959, there were only two beryllium suppliers in the United States, and only one of these, Brush Beryllium, had ever forged beryllium ingots of sufficient size and purity. In addition, making a forged, beryllium heat sink wasn't exactly cheap and the shields weighed over 350 lbs. apiece when they were done.

The second idea required the design and fabrication of an ablative fiberglass heat shield, one that would be partially consumed by the heat of reentry. In 1959, the idea of using an ablative heat shield was fairly new, the concept having been pioneered by Harry Julian Allen, a senior aeronautical engineer at the Ames Research Center in 1952.

Allen had done considerable work with ablative technology about seven years earlier and there was something very neat and tidy to the STG engineers about having a heat shield that rejected heat based on a physical transformation (i.e., from a solid into a gas); in other words, it took a lot more heat to change the physical state of a material than it did to raise temperature. In 1952 when Convair engineers were trying to develop a shape for Atlas missile nose cones (a shape that would have a nuclear weapon inside), using a digital computer, they initially concluded that a long needle-nose shape would be best. However, when they actually tested the shape in the supersonic wind tunnel at Ames, they discovered to their horror that the heat transfer from the exterior to the interior of the nose cone would turn the weapon into gas before it could be detonated. To quote Allen, the Convair engineers, "cut off their computer too soon." After making some rough "back of the envelope calculations," Allen discovered that the amount of heat absorbed by an object during reentry depended upon the ratio between pressure and skin drag. If someone used a blunt-nosed reentry vehicle, the unaerodynamic shape would generate considerable pressure drag at the nose of the craft; i.e., a shock wave would be created that would absorb much of the kinetic energy created during reentry. As Allen later recalled, "Half of the heat generated by friction was going into the missiles. . . I reasoned that we had to deflect the heat into the air and let it dissipate." Streamlining was out and blunt was in.

What NASA decided to do in the end was hedge their bets by manufacturing both beryllium heat sinks and laminated fiberglass heat shields and defer the decision until a later date. McDonnell Aircraft would ensure that any spacecraft they built could be fitted with either shield, However, following the Big Joe test flight on March 9, 1960, NASA no longer had any doubts about which way to go with respect to heat shields for orbital flight. What they did was mount a Mercury boilerplate spacecraft (fitted with an ablative heat shield) on top of an Atlas booster in order to determine how the spacecraft would perform during reentry into the Earth's atmosphere at orbital velocity, or 17,000 mph. This could only be done using an Atlas booster, as opposed to the smaller Little Joe launch vehicles fired from Wallops Island (the Little Joe was actually constructed from a cluster of Pollus and Recruit rockets). Even though the temperamental Atlas booster malfunctioned during staging (the sustainer engines got hung up in their racks and didn't jettison as designed), the resulting flight profile more than satisfied the mission objects and in certain aspects, exceeded them. The experimental ablative heat shield was subjected to more than 3,500 degrees of heating and the capsule hung together even after reentering the atmosphere at a high angle of attack, blunt end first, without the benefit of a stabilization control system.

When McDonnell submitted their preliminary design for the spacecraft, as an option to the tractor-style escape rocket, they also offered a fin-stabilized rocket escape system integrated into the bottom of the capsule, along with the retro and posigrade rockets. In fact, when an early flight test with the tractor type escape rocket failed, NASA seriously considered abandoning that approach.

This happened when the first pad abort flight tests at Wallops Island on March 11, 1959 were dismal failures. After a normal liftoff, the Recruit tractor rocket and capsule boilerplate ended up doing two complete loops before smashing into the beach at a phenomenal speed. Obviously, if it had been a full-up manned spacecraft, the astronaut would have been killed. However, after dragging the mangled wreckage out of the surf on the Virginia beach, they discovered that a graphite liner had blown out of one of the rocket's three exhaust nozzles. Thus, the fact that the test failed was not an indication that the concept was unfeasible. As a result, three STG engineers, Willard S. Blanchard, Jr., Sherwood Hoffman, and James R. Raper put long hours in over the next month to develop an improved escape rocket design. One thing they did was deliberately misalign the rocket with respect to the mounting pylon (the tubular structure that connected the rocket to the spacecraft). This guaranteed that when the escape rocket was ignited, it would pull the spacecraft to safety away from an exploding booster.

The Mercury escape tower was finally given a critical test during the Little Joe 2 launch from Wallops Island and this particular boilerplate also carried a small primate, Sam, an all-American Rhesus monkey. What this test was supposed to validate were the motions of the spacecraft-tower combination during a high altitude abort, the physiological effects of acceleration on a small primate, drogue chute operation, and general recovery procedures. The escape tower fired as planned 59 seconds after launch on December 4, 1959, taking the boilerplate and Sam for a ride up to about 53 miles. However, according to Space Task Group engineer Robert F. Thompson, recovering the floating capsule from the Atlantic Ocean was anything but easy:

> ... he [the Captain of the ship] let the ship slow down too much, and the ship... turned sideways into the waves... it started rolling forty-five degrees about the time we hooked onto this capsule. Well, one minute the capsule would be in the water, then another minute it'd be twenty feet in the air and swinging like a wrecking ball, but, anyway, it banged against the ship once or twice. Finally as it started swinging in, we said, "Turn it loose," and the sailors released the line... So we got back aboard the ship pretty quick, but just as we got it aboard ship, one of the sailors who was out on the deck got washed overboard. So we got the capsule here, but here's this kid in a life jacket just waving like mad, going up on the top of one wave and down another one.
> — Robert F. Thompson, Johnson Space Center Oral History Project
> August 29, 2000

Even after the harrowing experience of getting the capsule on board, Thompson and his crew were then faced with the difficult task of getting the monkey out of the capsule. Unlike the later production models, the Mercury boilerplate test vehicles did not even have hatches. They had to be disassembled into halves by removing over 70 bolts, not an easy task on a rolling ship at sea:

> *Most of the technicians that I had taken from Langley with me were seasick, so I had a bunch of sailors, and we'd never seen this damned capsule that the monkey was in (they had planned to hand the capsule over to the Air Force technicians). But in any event, we got these bolts off, lifted that afterbody and set it off and got this aluminum can that was about the size of a big waste basket, and the monkey was in there... it turned out they (the technicians) were all seasick over on the LSD. In fact, they were over there giving themselves the sugar solution that they were going to give the monkey to keep themselves from dehydrating...*
> — Robert F. Thompson, Johnson Space Center Oral History Project
> August 29, 2000

The final escape rocket system design used a single solid propellant rocket with three exhaust nozzles canted 19 degrees off centerline having a resultant thrust of 52,000 lbs. (the escape rocket was also used to jettison the pylon assembly during a normal launch sequence). When you think about it, the Project Mercury escape rocket system was one of the best features of the whole concept. Why? Because it gave a warm and fuzzy to any astronaut that if the worst happened, he had a way to get his precious body off the launch pad and away from what could be a very bad place indeed. To initiate an abort sequence, the astronaut had to do two things: first, he used his left thumb to depress a small mechanical release button on the top of the abort handle, mounted to the left of the couch. This button unlocked the handle, allowing it to be swung or rotated outboard, actuating the abort switch. From then on, the pilot's life was in the hands of the Lockheed Propulsion Company of Redlands, California and polysulfide ammonium perchlorate (a Class B Explosive), the chemical formula used to power the escape rocket. The sequence of events during a Mercury abort were quick and violent. First, the capsule adapter clamp ring was released by firing several small explosive bolts, mechanically freeing the capsule from the rocket, whether it be an exploding Redstone or one of the many Atlas missiles that self-destructed on a regular basis. A matter of milliseconds later, the Lockheed escape tower rocket ignited with a roar tearing the capsule away from the booster at a rate of acceleration of up to 20 g's. This carried the capsule and astronaut up to an altitude of about 2,500 feet (assuming it was a launch pad abort). Next, the capsule tower clamp ring was released (this is the same ring we used to lift Liberty Bell 7 in 1999) and a small Atlantic Research tower jettison rocket ignited, carrying the expended escape rocket and support structure away. Two seconds later, the capsule's drogue parachute deployed and a couple of seconds after that, the antenna fairing was history. Twelve seconds later and only 16 seconds after reaching altitude, the impact landing skirt was deployed. When all was said and done, less than 30 seconds after activation, the sequence was complete and the astronaut, hopefully, still alive and without serious injuries. While the Mercury escape tower was never used on an actual manned mission, Soviet cosmonauts Titov and Strekalov did have the opportunity to experience a launch pad abort on September 27, 1983 during the attempted launch of Soyuz T-10. At T minus 90 seconds, a large fire started at the base of the launch vehicle. Twelve very long seconds later after the flames had burned through the wires controlling the automatic abort sequence, the launch controllers ignited their escape tower (it was a copy of Faget's design) which quickly dragged the two men to a safe landing about 2.5 miles away from the conflagration. In spite of having been subjected to an acceleration of up to 17 g's, both men escaped unharmed from their flight lasting five minutes and thirty seconds. Years later, both men thanked Faget in person for creating the escape system that saved their lives.

Another general design issue needing resolution was how to slow the spacecraft down after reentry so the astronaut could land (on the ocean or dry land) without getting killed. In fact, many of the engineers had to redirect their thinking, away from the narrow streamlined shapes they were used to and towards the blunt reentry vehicle the Mercury spacecraft was destined to be (with respect to aerodynamic stability). To this end, they conducted a series of wind-tunnel experiments at Langley and the Ames Research Center to establish the stability of Faget's cone shaped contraption at high mach numbers. One of the great things about his design was that the Mercury spacecraft was passively stable, from the aerodynamic standpoint. In other words, no matter what happened, once the capsule started encountering the fingers of the upper atmosphere, it would reenter with the heat shield pointed towards the direction of flight. There were two reasons for this: First of all, while the capsule's center of gravity was centered with respect to the pitch and roll axes, in the longitudinal direction (i.e., the flight path), the C.G. was very close to the heat shield, actually slightly below the edge of the lower explosive hatch sill. It was sort of like having a child's' top with a bottom filled with lead; it was hard to turn over onto its pointed end. In addition, in the final spacecraft design, the antenna canister on the capsule's small end was fitted with a "destabilization flap," a small metal spring-loaded control flap that made it so that no matter which way the spacecraft encountered the atmosphere, it would automatically reorient itself with the heat shield pointed forward, in what NASA called the "reentry attitude." In fact, these two design features may have saved astronaut Scott Carpenter's life during his Mercury mission after he expended all of his attitude control fuel during reentry.

Once NASA established the general stability of the capsule design, they had to figure out a way to slow it down to a suitable landing velocity without shredding the parachutes. They had concluded earlier that the ideal landing speed should be no more than 30 feet per second. Surprisingly, very little was known in 1959 about the behavior of parachutes at high altitude. What they discovered was that high altitude and high speed parachute deployment might cause what was called "squidding," where the chute never really fully opened. This problem was tackled by engineers from the US Air Force, Langley, McDonnell, and Northrop's Radioplane division, the subcontractor responsible for much of the recovery equipment on Project Mercury (Radioplane also supplied the dreaded SOFAR bombs, later to become a significant safety issue during Liberty Bell 7's recovery in 1999). Ultimately, they decided to use a 63-foot diameter "ringsail" parachute for Project Mercury. In the final capsule design, there would actually be four parachutes fitted to Liberty Bell 7: First, a drogue chute that would be deployed at around 20,000 feet to initially slow the capsule down. Secondly, a ringsail parachute that would decelerate the capsule to landing velocity. The ringsail chute would not open fully when it was first deployed. In order to reduce the opening shock, it was first deployed in a "reefed" condition, where the parachute's full diameter was restricted until the capsule had slowed to a predetermined velocity. Once that had happened, a series of "reefing line cutters" fired, allowing the parachute to deploy more fully and slow the spacecraft down to the targeted 30 foot per second impact speed. The third parachute was a reserve parachute identical to the ringsail chute. Finally, if all else failed, the astronaut could attach a small personal parachute to his space suit harness and bail out of the spacecraft, assuming he could even get clear of the thing as it fell at high speed towards the ocean. This final, last ditch parachute was suggested by astronaut Alan Shepard and was mounted just under the edge of the capsule's hatch sill (the personal parachutes also incorporated reefing line cutters, making it necessary to partially deploy Grissom's chute on the deck of the recovery ship in 1999 to remove the small, but still dangerous, pyrotechnic devices). It's easy to imagine that any astronaut could get out of a Mercury spacecraft if he had to, no matter how hard it might be. Survival is a great motivator.

The engineers at the Space Task Group were also concerned about the g forces the astronaut would experience when the spacecraft actually touched down. This was a serious concern during an aborted flight, as there was the possibility that a Mercury capsule might end up on dry land where the ground impact forces were far greater than in a water landing. To this end, NASA initiated their infamous "pig drop" experiments. This was where they strapped four large (and no doubt terrified) Yorkshire pigs into specially fitted contour couches and boilerplate capsule structures and dropped them from varying heights to determine whether or not a human could survive similar decelerations. The experiments tested both the concept of the Mercury contour couch as well as the aluminum honeycomb energy-absorption system, and once again, Caldwell C. Johnson developed the requisite detailed sketches showing the hapless swine strapped inside of a boilerplate capsule. The pig drop tests were actually quite an attention getter at McDonnell Aircraft as many employees abandoned their work stations to see the pigs hit bottom, so to speak. While the fearless swine (a pig named "Gentle Bess" was the prime candidate) endured decelerations of 38 to 58 g's before experiencing minor internal injuries, it was probably more impressive to the test personnel to see them get up and walk away after the event. Whether the pigs ended up on McDonnell dinner plates or were released unharmed for their courage is unknown. The important thing was that NASA was able to establish that an astronaut could survive an impact on dry land.

By mid-March of 1959, McDonnell had displayed their first full scale capsule mockup to the Space Task Group for review, as well as a detailed 80 page set of specifications with a written explanation of each primary subsystem, such as the adapter ring (to the booster), retrorocket package, heat shield, pressure bulkheads, structure, antenna canister, and escape rocket pylon. This mockup was used by NASA engineers to study critical design features such as the accessibility to controls, and astronaut egress and escape.

From the airframe standpoint, what McDonnell eventually created was a marvel in 1960's materials technology. Overall, the basic physical structure of the spacecraft was made from a framework of longitudinal "hat stringers," each made of a titanium alloy containing aluminum and tin. These stringers, coupled with framing rings made from an aluminum-vanadium alloy of titanium, were welded together and formed the basic cone shape of the spacecraft. All of the welded components were made from annealed materials, so no strength was lost during the later spot welding process.

The panels forming the pressurized cabin were created from commercially pure titanium and were of a double wall construction. While the inner sheeting was flat, the outer panels were formed from 0.010 inch thick titanium in a drop hammer using Kirksite dies (this outer section was beaded or corrugated, giving the composite wall section a rigidity equivalent to a much thicker structure). These two skins were fusion-welded together in an inert gas atmosphere; an impressive accomplishment, considering the minimal thickness of the materials being used. The pressure skin and capsule framework were then spot welded together (additional thickness was added around the hatch and window openings). In some cases, as many as seven thicknesses were joined during the spot welding process. In addition, each pressure skin corrugation was individually sealed by welding, further stiffening the structure and effectively turning the space between the two walls into hundreds of separate compartments. It took over 20,500 inches of seam welding (done on Sciaky three-phase welders with Dekatron controls) to complete each spacecraft. All of the above procedures created the overall capsule structure and astronaut cabin, allowing the astronaut to exist in a pure oxygen atmosphere holding a nominal pressure of 5.5 psi. While from the structural standpoint, the basic Liberty Bell 7 assembly was very rigid in the longitudinal axis, it was less so with respect to side loadings. This fact was borne out by the fact that one of Liberty Bell 7's titanium frames was slightly creased, the damage probably caused by the capsule's lateral impact with the ocean after the craft was released by the recovery helicopter.

However, this basic assembly of the spacecraft was not very resistant to the predicted temperature extremes of space flight; for that, they added the familiar dull black corrugated shingles on the capsule's exterior. While the exterior corrugated shingles appeared to be painted flat black, in fact, they were not painted at all. The shingles, manufactured from a heat resistant nickel-base alloy (having a thickness of 0.016 inches), René 41, were actually heat treated to develop a black oxide layer having high thermal emissivity. They were formed to contour from annealed stock at room temperature, again employing a drop hammer and Kirksite dies. McDonnell was very careful to do this particular heat treating process right, as small amounts of contaminants could form spots of green oxide, potentially causing hot spots, which had to be manually removed using wire brushing, vapor blasting, and pickling. If that happened, they had to do the process all over again. When the shingles were mounted to the capsule's titanium structure, McDonnell used oversized holes to allow for free expansion during the blistering heat of reentry. What's interesting is that following Liberty Bell 7's recovery and during the disassembly process, every single one of the stainless-steel fasteners connecting the shingles to the titanium structure was easily removed following a slight strike with a Phillips head screwdriver and hammer. In fact, each one of these fasteners was individually fitted to its respective hole to guarantee the appropriate mechanical clearances. Finally, thermoflex insulation was added to the vacant area in between the longitudinal stringers and Min-K insulation at the point of contact underneath the exterior shingles to reduce heat transfer.

A new addition to spacecraft No. 11, Liberty Bell 7, was a large observation window integrated into the pressure cabin and positioned so the pilot could observe the capsule's attitude visually in addition to using the attitude control display. The new window consisted of an outer, curved pane (0.350 inch vycor glass) and an inner flat triple pane glass (each, a separate pane of 0.340 inch thickness) structure designed to withstand nominal cabin pressure. The inner surface of the outer window pane and both surfaces of the inner window assembly were coated with a single layer of magnesium fluoride (M_gF_2) film to help impede thermal radiation into the cabin.

The astronaut could also observe the Earth by using an optical periscope, a navigational instrument integrated into the capsule's structure. The device, manufactured by Perkin Elmer, was made up of three major assemblies: the display, and upper and lower housing assemblies. The periscope display allowed the pilot to determine his drift, altitude, pitch, roll, true vertical, retro attitude, and relative bearing to the Sun and Moon (all of it with respect to the capsule's position relative to the Earth). Integrated into this circular display were all of the scales, indicators, reticles, etc. needed for the pilot to both observe the Earth and determine whether or not his flight attitude was correct for that particular phase of the flight. The periscope's upper housing assembly contained assorted mirrors and filters and formed the base for the field lenses, which are what actually projected the Earth's image onto the display assembly. The instrument's lower housing contained more lens components which could be used to magnify the image.

The periscope was rated for loads up to 100 g's (interesting, as such forces could easily kill any astronaut needing to use the instrument) and was designed to be extended or retracted, depending upon what phase of the flight was in progress. With the spacecraft and booster on the launch pad, the periscope was extended (the area surrounding the extended objective lens was also used to house an electrical connector which fed electrical power to the capsule before launch – this is where the launch umbilical was attached). At launch, as the umbilical disconnected, the periscope retracted and remained so until capsule separation, after which time it was extended for use during flight above the Earth. Thirty seconds after the retro package was jettisoned, the periscope was again retracted until an altitude of about 10,000 feet, after which it was extended for the last time (all of these procedures could be done manually as well). The astronaut could even observe the surface of the ocean as the capsule descended on its main parachutes.

For thermal protection during reentry, the two suborbital spacecraft (Freedom 7 and Liberty Bell 7) were fitted with beryllium heat sinks, as opposed to the ablative heat shields used on the spacecraft configured for orbital flight. The expensive heat sinks weighed 342 lbs. in air, were 74.5" in diameter with a spherical radius of 80 inches, and forged from hot pressed sintered QMV grade beryllium. However, during reentry, it was expected that the exterior of the parachute compartment would also be subjected to considerable thermal heating. As a result, on the orbital craft, this area was also covered with beryllium panels due to that material's high specific heat, low density, and low oxidation rate (they were covered on the interior with a thermal conversion coating of gold, along with an oxide barrier; the panel's exterior was treated with a black ceramic paint). Shepard's and Grissom's craft, however, used 0.22 inch thick aluminum panels in this area as the heat generated by a suborbital mission was expected to be considerably less.

The capsule's interior brackets and frames were of conventional aircraft construction and made from high-strength aluminum alloy, along with some magnesium and steel alloys, where needed. Most of the high pressure gas tubing was stainless steel with the hydrogen peroxide lines for the reaction control system made of welded aluminum tubing. The all important escape tower pylon was a welded construction of normalized 4130 steel with all welded joints being stress-relieved (probably by shot-peening) for maximum strength.

Finally, the astronaut's couch was made from a sandwich structure, consisting of an aluminum honeycomb core with faces made of fiberglass. Into this basic structure, a liner conforming to each individual astronaut was fitted, covered with a thin sheet of foam rubber for comfort. For added impact protection, more aluminum honeycomb material was mounted below the couch. Once all of the astronaut's individual couches were completed, the pilot's names were stenciled on the back. Surprisingly, according to Guenter Wendt, these names were in many cases removed before flight so the Soviets could not find out who was scheduled for which mission.

One of the reasons that McDonnell was able to get away with such a light structure for a pressurized cabin was NASA's early decision in favor of a single gas oxygen-based environmental control system. While some officials advocated a true shirtsleeve environment using a two-gas system of oxygen and nitrogen at sea level pressure (i.e., 14.7 psi), the increase in weight to hold that much pressure in the vacuum of space was prohibitive. Also, a mixed gas environmental control system introduced considerably more complexity into the spacecraft than needed, according to Max Faget:

> *The most important consideration in choice of a single gas atmosphere is reliability of operation. If a mixed gas atmosphere were used, a major increase in complexity in the atmospheric control system and in monitoring and display instrumentation would have resulted. Furthermore, the use of a mixed gas system would have precluded the use of simple mechanical systems for a great number of these functions which in itself would have decreased the reliability of performance.*
> — Dr. Max Faget, NASA Space Task Group

"Reliability of performance" was a major issue because as of January 1960, neither the cabin or suit environmental control systems had passed their test to operate as designed for 28 hours.

The whole concept of a single gas system is related to partial pressures and the efficiency of the human lung to absorb oxygen. At one atmosphere, or a sea level pressure of slightly less than 15 psi, in an atmosphere of 80% nitrogen (an inert gas) and roughly 20% oxygen, the human body receives sufficient oxygen to maintain life. Unfortunately, the amount of structure to hold that much pressure in a vacuum is considerable. On the other hand, with a spacecraft structure designed to hold only 5 psi while using a 100% oxygen environment, the astronaut will still receive sufficient oxygen to live. This is because the number of oxygen molecules in 15 psi of air (80% nitrogen and 20% oxygen) is comparable to the number of oxygen molecules in pure oxygen at only 5 psi. In a nutshell, using a single gas atmosphere was a more efficient way to get the astronaut the amount of oxygen he needed to live. Unfortunately, while they didn't have any problems with the concept in Projects Mercury and Gemini, NASA learned the hard way with Apollo I that pure oxygen on the launch pad is also one hell of a fire hazard.

To make this single gas system work, McDonnell turned to AiResearch, the industry leaders in such technology, who ultimately developed a Rube Goldberg contraption that fit neatly underneath the astronaut's couch. At the heart of the system were three 7,500 psi titanium oxygen tanks about the size of grapefruit, coupled with pressure reducing valves and regulators to supply the precious gas at a reduced pressure (two tanks were for breathing gas and the third pressurized a water coolant tank). Living within this maze of piping, compressors, a solids trap, butterfly valves, pressure sensors, a heat exchanger, and assorted plumbing were also a lithium hydroxide canister (to remove carbon dioxide and odors from the cabin and space suit), blowers, and plumbing for a fresh air snorkel to input atmospheric air after the parachutes had deployed and the spacecraft was on its way down. The ingenious system also used a mechanism to squeeze a small sponge to remove perspiration and water vapor from the astronaut's breathing loop, as well as a water coolant tank to help cool the cabin interior (the coolant tank, according to NASA, could also be used as a source of drinking water, assuming anyone wanted to drink the stuff to begin with). The system even had the ability to be charged with Freon before launch to keep the astronaut cool before liftoff. Once in flight, heat could also be supplied via the above system by circulating the coolant tank water through three large (and warm) electrical inverters. The overall temperature in the craft's cabin was controlled by a simple selector valve, which regulated the amount of water entering the cabin heat exchanger.

Both the Mercury-Atlas 1 (MA-1) and Mercury-Redstone 1 (MR-1) launches in 1960 represented low points with respect to NASA's bid to get an American into orbit as they were dismal, and very public, failures. What they hoped to do on the MA-1 flight was to qualify the integrity of the capsule structure and afterbody shingles under the expected loads of a typical Mercury mission. The massive Atlas booster, constructed much like an inflatable stainless-steel balloon, lifted off from the pad on July 29th. However, only a minute after Atlas missile No. 50-D punched through the Florida cloud cover, all hell broke loose:

> *About one minute after liftoff all contact with the Atlas was lost. This included telemetry and all beacons and transponders. About one second before telemetry was lost, the pressure difference between the lox and fuel tanks suddenly went to zero...*
>
> — NASA Space Task Group, 1960

Whatever occurred, happened very quickly as the Atlas broke up at maximum dynamic pressure (i.e., max q) and what's worse, this particular flight did not use an escape tower for the capsule. As a result, the whole assembly disintegrated in flight as pieces of spacecraft, structure, and rocket engines, peppered the surface of the Atlantic Ocean. A later analysis of the recovered wreckage indicated a possible failure of the structure making up the capsule to booster adapter. Much of the spacecraft's mangled wreckage was recovered from the shallow 40 foot deep waters about seven miles off the coast of Florida.

What finally happened to the MA-1 capsule wreckage is a story in itself. Apparently, at some point following the mission (probably years later), the spacecraft debris was spotted at a US government auction by someone who recognized it for what it was (i.e., Mercury spacecraft components). The capsule wreckage was purchased and ended up sitting in a backyard for almost 20 years, mostly doing nothing but collecting leaves. It was rediscovered by Max Ary, the President of the Kansas Cosmosphere and Space Center, who managed to negotiate some sort of deal with the owner, after which, it was trucked to Ary's Hutchinson, Kansas museum. Not much was left, except for most of the large pressure bulkhead and sections of the afterbody and forebody areas. This wreckage was used in 1992 to confirm the fit and function of the recovery tools used to eventually raise Liberty Bell 7 in 1999, as well as for general analysis to determine the materials used to build the spacecraft.

As with MA-1, the first attempt to launch MR-1 was a public embarrassment for the fledgling space agency. However, in this incident, they at least managed to save the booster and capsule for a later, and successful, test flight. On November 21, 1961 at 9:00 a.m., the firing command signal was given for a flight that would hopefully qualify the capsule's automatic stabilization and control system, reaction control system, posigrade rockets, and recovery system. Unfortunately, after the Redstone booster ignited and rose only four inches off the launch pad, the booster's Rocketdyne A-7 engine immediately shut down causing the booster / spacecraft stack to settle back down on the pad, fortunately, without tipping over. Shocked NASA engineers watched helplessly as the capsule's escape tower jettison rocket fired, sending the steel pylon structure to a landing about 400 yards from the pad. If that wasn't bad enough, three seconds later the capsule's drogue chute popped out of the top of the antenna canister, followed by the main parachute shortly thereafter.

Now they had a real problem because sitting on the ground was a fully fueled Redstone booster with an armed destruct system and a spacecraft with live posigrade and retrorockets – all of it sitting there connected to a 63 foot diameter ringsail parachute threatening to topple the whole thing over at the slightest gust of wind. According to Guenter Wendt, the McDonnell Aircraft pad leader, Kurt Debus, the Launch Director, suggested that they get one of the security guards to put a 30.06 bullet through the rocket to help relieve rising pressure on the Redstone's main tank. Both John Yardley and Chris Kraft nixed this idea. Shooting the booster full of holes could easily create the massive explosion they were hoping to avoid. Eventually, a mechanic volunteered to crawl underneath the missile and he hooked up a small gas line to the rocket which popped a small relief valve, avoiding the immediate danger. In the end Wendt, along with Bob Jones and Bob Graham, courageously rode the elevator to the top of the gantry and very carefully disarmed the active systems. An investigation eventually discovered that the rocket's sequencing was thrown off when a two-pronged electrical plug disconnected without the booster being grounded, causing the engine to shut down. Even though MR-1 did not do what it was supposed to, the booster and spacecraft were successfully launched on a follow up mission, MR-1A, on December 19th, enabling NASA to qualify the all important Automatic Stabilization Control System.

The spacecraft's Automatic Stabilization Control System (ASCS) was not one subsystem, but actually an integrated combination of sensors (i.e., horizon scanners), gyros, rate transducers, and the capsule's automatic Reaction Control System (RCS). There were two horizon scanners fitted to all Mercury spacecraft, both integrated into the capsule's antenna canisters: one scanner detected the capsule's attitude in the pitch axis (it was mounted on the very end of the canister) and the other established spacecraft orientation in the roll axis (fixed to the canister's side). Both scanners sensed the difference between the amount of infrared radiation received from the surface of the Earth as compared to the zero radiation from space. This sensory data was fed to the capsule's attitude rate indicator (which displayed Earth referenced attitude and rates of pitch, roll, and yaw) and was also used to control the capsule in four basic automatic flight sequence modes: Damper, Orientation, Attitude Hold, and Reentry. These flight modes were actually sequences of capsule maneuvers where the spacecraft was commanded by the ASCS to do certain things depending upon mission timing and which phase of the flight was in progress at the time. The "Damper" sequence mode started just after capsule separation from the booster and provided rate damping to minimize any tumbling action. The "Orientation" mode turned the spacecraft around (a 180 degree clockwise yaw rotation) and also pitched the capsule down into the "retrograde firing angle," which meant a 34 degree nose down attitude, a safe orientation for firing the retro rockets if the mission had to be quickly aborted (this happened within 30 seconds of the start of the orientation mode). "Attitude Hold" did just that – it continued to hold the spacecraft with the nose down in the retro attitude, which was the normal flight orientation when in orbit. At this time, the pilot could assume manual control of the spacecraft using the Fly-By-Wire mode and any maneuvers could be assisted with an automatic rate damping feature. Finally, the "Reentry Mode" ensured that the capsule was in the correct attitude for retrofire, held the spacecraft as the retro rockets fired and after the retropack was jettisoned, and initiated more rate damping and a 10 to 12 degree/second roll rate during atmospheric reentry. If everything was working properly and the astronaut left the automatic system engaged, the spacecraft could theoretically fly itself from launch to landing (the hand controller was even fitted with a safety pin so it could be immobilized, if desired).

If the astronaut grew bored at the prospect of being chauffeured through space (a lá "Spam in the can"), he could disengage the automatic system and control the capsule manually using either the Fly-by-Wire or Rate Command Systems. While both the Fly-by-Wire and Rate Command systems allowed the astronaut to fly the capsule using the hand controller, the Fly-by-Wire mode used the Automatic Stabilization Control System while the Rate Command mode used the Rate Stabilization Control System (RSCS). The ASCS and RSCS were totally independent of each other and even used different sets of thrusters. These two flight control systems, while similar in that they used the astronaut's hand controller, functioned in very different ways. In the Fly-By-Wire mode, moving the astronaut's hand controller (i.e., joystick) closed any one of several small microswitches, depending upon the amount of stick movement. In other words, moving the stick 30% or 75% of full travel resulted in either low or high thrust actuation, respectively. This way, the pilot could fly the capsule with either fine or coarse control as the situation dictated. This was a totally electric control system, as the nomenclature "Fly-by-Wire" suggests. Using the Rate Command Mode meant that instead of closing small electrical switches, control stick movement instead manually actuated proportional fuel control valves (via a control linkage assembly) which resulted in proportional rates of movement. In addition, in this rate mode, stick position data was fed back to the capsule's control system which compared control stick position with actual rates of movement. If the commanded rate did not match what was actually happening, electrically controlled solenoid control valves in the manual Reaction Control System were also actuated, helping things out, so to speak. One way or the other, the net result was that the rate system caused steady state angular rates of spacecraft movement. As a result, a greater angle of control stick deflection meant a faster rate of pitch, roll, or yaw, depending upon which direction the stick was moved in. Pushing the stick forward and backwards created movements in the pitch (or up/down axis), moving it left and right caused the capsule to roll counterclockwise or clockwise, and twisting the stick to the left of right made the capsule's nose swing to the left and right respectively.

All of the above control systems were coupled with flight control hardware, i.e., or the mechanical and fluid components used to orient the Mercury capsule as it floated high above the Earth. Mercury spacecraft had two separate and redundant sets of Reaction Control Systems (RCS), automatic and manual, the former designed to work with the ASCS and the latter working with the RSCS. Both systems incorporated high strength hydrogen peroxide as a fuel (90% purity H_2O_2), an inert gas (helium) pressurization system, related valving, and a series of reaction control thrusters.

The theory behind using hydrogen peroxide as a fuel was simple: according to the Project Mercury Familiarization Manual, one pound of 90% purity hydrogen peroxide solution, when catalytically decomposed, produces about 60 cubic feet of gas (actually a combination of water vapor, oxygen, and heat) that can be used to control the movements of the spacecraft. On a Mercury spacecraft, the fuel for both the automatic and manual systems was contained inside of two separate flexible torus bladders in the area between the heat shield and aft pressure bulkhead. These bladders, when externally pressurized with helium (they were mounted within small tubular tanks), fed the caustic fuel into the spacecraft control systems to be distributed as needed, depending upon which flight mode was in operation. When the liquid fuel entered one of the reaction control nozzles through a small metering orifice, it was decomposed over a catalyst bed (a stack of nickel screens, coated with silver and drexite), where a violent chemical reaction generated gas, which exploded through a small plenum chamber and out the nozzle. Exhaust gas temperatures could be as much as 1,400 degrees F and these thrust nozzles created anywhere from 1 to 24 lbs. of thrust. The Automatic RCS used twelve separate nozzles (four per axis) for all three axis of control: 24 and 1 lb. thrust nozzles for the pitch and yaw axis and 6 and 1 lb. thrust nozzles for the roll axis. On the other hand, the Manual RCS used only six separate thrust chambers: 0 to 24 lb. nozzles for the pitch and yaw axis and 0 to 6 lb. nozzles for the roll axis. All of these fuel systems could be manually activated or disabled and dumping the hydrogen peroxide fuel prior to landing in the ocean was a standard procedure on Mercury missions. Typically, the astronaut used either the Automatic or Manual RCS, not both at the same time (though this could be done) as the Manual System (with the astronaut in the loop) was intended as a back-up to the Automatic system. Unfortunately, when Scott Carpenter was trying to orient his Aurora 7 spacecraft for reentry during his May 1962 mission, he accidentally left both systems on, causing precious attitude control fuel to be used at an alarming rate. There is a photograph taken of John Glenn embracing Scott Carpenter after his mission, not surprising as they reportedly became close friends during Project Mercury. Some feel that Carpenter was lucky to survive reentry. It is possible that, from Glenn's perspective, Carpenter had risen from the dead when he made it back instead of being fried during reentry.

In addition to the hydrogen peroxide-based rocket system, Mercury spacecraft also had a series of solid fuel rockets making up the posigrade and retrograde rocket systems. Each had a separate and unique function. The purpose of the posigrade rocket system was to accomplish separation between the capsule and booster at a rate of 15 feet per second once orbital velocity had been achieved. In the case of a suborbital mission, the three posigrade (all fired simultaneously) rockets simply helped increase the distance between the capsule and booster once they had separated. Each posigrade, actually an Atlas retro-rocket, used Arcite 377 for propellant and developed a thrust of about 370 lbs. The system was triple redundant in that only one rocket was needed for proper separation.

The retrograde rocket system was designed to slow an orbital spacecraft down sufficiently to allow it to reenter the atmosphere in a controlled manner and towards a planned landing point. The three retro rockets, mounted together with the posigrades in the retrograde package, were ripple fired in that they were ignited in sequence at five second intervals. Each individual rocket nozzle was fitted with a protective cover to protect it from micro-meteorites, had a nominal thrust of 992 lbs. and a burning time of 13.2 seconds. The rockets were also heated using thermal blankets and fitted with dual igniters to guarantee as much reliability as possible. Once the retro rockets had fired out, the entire package was jettisoned 60 seconds later (with the assistance of a coil spring to eject the package from the capsule) by firing three explosive bolts which in turn released the three restraint straps. As with the posigrade system, the system was triple redundant in that only one of the three rockets was actually needed to deorbit the spacecraft.

Many of the spacecraft's Environmental Control System components were qualified during one of NASA's most important early test flights, Mercury Redstone No. 2 (MR-2), which also had on board "Ham," a 44 month old, 37 lb. trained Chimpanzee. These particular animals were invaluable in gathering important aeromedical data, as well as testing all of the telemetry equipment needed to transmit flight data from the spacecraft to the ground. In addition, Chimpanzees by nature respond to physical stimuli in a manner similar to that of humans and also have a nearly identical anatomy, from the standpoint of organ placement and internal suspension. Ham was selected for his historic mission only a day before flight out of a colony of six chimps (a female was the alternate) and tasked to pull specific levers when ordered by a small flashing light. To not obey, meant that the astro-chimp would be given a small, but uncomfortable, electric shock on the soles of his feet. From Ham's perspective, it was important to not get shocked.

Just before noon on January 31, 1961, Ham, Mercury spacecraft no. 5, and the Redstone booster were launched from Cape Canaveral. About a minute into the flight, NASA computers sensed that the rocket's trajectory was too high and increasing and soon started predicting a 17 g load during reentry, considerably higher than planned. This was caused by a higher than nominal thrust with the corresponding increase in velocity (7,540 feet per second as opposed to the planned 6,565). What's worse is that the spacecraft's internal pressure plummeted from 5 down to only 1 psi, a failure later traced to a malfunction in the capsule's air inlet snorkel valve.

On the good side, in spite of the abnormal flight profile, the capsule's escape tower fired as planned, though slightly early, thus confirming that system's function during an abort scenario. However, the net result was that Ham landed 48 miles farther downrange than expected:

> ... the Redstone integrating velocity meter didn't work just right, so the Redstone didn't give the spacecraft the cut-off signal it was supposed to get. When it burned out, the spacecraft thought it was an abort condition, so it fired the retrorocket. So Ham got the planned boost out of the Redstone plus the additional boost out of the retro rocket fired when it wasn't planned to fire, which sent him about, I don't know, maybe 100 miles farther than he was supposed to go.
> — Robert F. Thompson, Johnson Space Center Oral History Project
> August 29, 2000

The worst part was that when recovery forces finally found the spacecraft, it was shipping water badly, laying on it's side, and in serious danger of sinking. This was caused when the beryllium heat sink (hanging below the capsule on the untested and flexible impact landing skirt) bashed the underside of the capsule, puncturing the titanium bulkhead. In addition, once the spacecraft tilted over enough, even more water started leaking through the defective snorkel valve (the beryllium heat sink eventually fell off the capsule):

> ... we quickly learned that that bag wasn't strong enough. We put the capsule with the bag the way we flew it in the tank at Langley, hit it with waves for about twenty minutes, and it just tore all to pieces in twenty minutes. So we put a whole bunch of steel cables in to reinforce the strength between the heat shield and the bottom bar of the capsule once it was deployed...We also then put a whole bunch of aluminum honeycomb on the bottom of this capsule so that when the heat shield hit there, it would hit honeycomb and not knock a hole in the bottom of the capsule.
> — Robert F. Thompson, Johnson Space Center Oral History Project
> August 29, 2000

Mercury Redstone No. 2 had a few problems, but it was a success in most respects and set the stage for the later manned suborbital missions. And Ham? After his historic space mission as the first "free" animal in space, he lived in the National Zoo in Washington, DC, for 17 years. After complaints from animal activists, in 1981, Ham was moved to a zoo in North Carolina where he was able to socialize with other chimps. After he died of old age on January 19, 1983, at age 27, Ham's body was shipped west, and later buried in the front lawn of the International Space Hall of Fame in Alamogordo, New Mexico.

Even though McDonnell's titanium cabin was designed to maintain an ambient pressure of about 5 psi in a "shirtsleeve" environment, the astronaut obviously needed a backup, i.e., some sort of pressurized space suit. Mercury astronaut Wally Schirra was assigned the responsibility of helping develop the space suit, in conjunction with the BF Goodrich company; ironic, because being inside of a pressurized space suit is not unlike living inside of an inflated automobile inner tube. In a close competition with the David Clark Company and the International Latex Corp., Goodrich won out in the end, probably due to their long experience with supplying the Navy with similar garments and previous history with helping aviator Wiley Post develop the first high-altitude pressurized flying suit.

One of the people instrumental in supplying Post with his "Winnie Mae" (Post's record-breaking aircraft) suit was Goodrich's Russell M. Colley, who worked with the famed pilot during the design of a crude aluminum helmet and rubber garment, reportedly stitched together on his wife's sewing machine. Colley later developed special swivel joints for the Goodrich suits supplied to the Naval Bureau of Aeronautics and together with Carl F. Effler, modified the standard U.S. Navy Mark IV pressure suit into what eventually became the silver colored suit made famous by Project Mercury.

One of the problems with the Mercury suits was that after they had been inflated a few times, the suit material stretched. This problem was solved through the use of improved textiles and helped maintain the suit's fit over time. In addition, there were numerous minor problems with zippers, laces, valves, gloves, etc., which were eventually rectified by redesigns or improved quality control. In fact, the space suit worn by Grissom during his Liberty Bell 7 mission was considerably improved over the one used by Shepard as it was fitted with the new rotating wrist bearings, allowing greater hand mobility. The suits were eventually given even more mobility when some thermal insulation was removed on the basis of data received during the Big Joe reentry heating test in September of 1959. The final design was silver in color, sealed with a zipper wrapping around the astronaut's torso, and fitted with a white space helmet coupled to a harness designed to keep the suit from ballooning when inflated. When worn by the appropriate steely-eyed missile man, the suit gave the astronaut the appearance of a gladiator from outer space.

To keep the astronaut in place during any aspect of space flight, planned or not, NASA and McDonnell developed an elaborate restraint system, drawing mostly on what was typically used on high performance test aircraft. This restraint system consisted of shoulder and chest straps, leg straps, a crotch strap, a lap belt, leg restraints, and if that wasn't enough, even toe guards. In fact, if the entire restraint system was used, the only part of his body the astronaut could really move were his arms. However, the leg straps were discontinued after Shepard's flight in 1961 and for comfort's sake, a harness reel control handle was mounted above and to the left of the astronaut's head to allow him to unlock and lock parts of the restraints as desired.

After Ham's historic flight, NASA continued working towards the objective of getting an American into space. However, on April 12th, 1961, at least part of the space race was over when the official Soviet news agency, TASS, transmitted the following translated press release:

> The world's first space ship Vostok with a man on board, has been launched on April 12 in the Soviet Union on a round-the-Earth orbit. The first space navigator is Soviet citizen Maj. Yuri Alekseyevich Gagarin. Bilateral radio communication has been established and is maintained with Gagarin.
> — Soviet News Agency TASS, April 12, 1961

Gagarin, riding inside of the Soviet twin module spacecraft, Korabl Sputnik VI (also known as Vostok I), call sign Swallow, was launched at 09:07 hours Moscow time on April 12th from the Baikonur Cosmodrome, near Tyura Tam, using the SL-3 variant of the SS-6 Sapwood rocket. The missile had three stages, the first stage being four external breakaway boosters strapped onto the second and third stages. The Soviet's Vostok was almost three times heavier than a Mercury spacecraft and used a sea level pressure mixed gas atmosphere and separable instrument section and retrorocket package for communications. The 27 year old Gagarin was the first person in history to be accelerated to a velocity of 17,400 miles per hour, endured 89 minutes of weightlessness, completed one near Polar orbit, and landed in good health near the Volga River, about 15 miles south of the city of Saratov.

Unknown to NASA was the fact that Gagarin did not ride his Vostok I spacecraft during the whole mission. At the time, the Soviets did not have a provision for softening the spacecraft's impact on dry land, even with the parachutes they used (all Soviet spacecraft are guided to impacts within their territorial boundaries). As a result, Gagarin ejected from his spacecraft at an altitude above 20,000 feet landing some distance from his craft. The spaceman was greeted after landing by an old woman, her granddaughter, and a cow. The Soviets hid the fact that Gagarin ejected from his craft because had that been known, it would have disqualified the cosmonaut from being awarded the international aviation altitude record which requires the pilot to remain with his craft until landing.

While there were similarities to what NASA hoped to do with Project Mercury, there were also significant differences. First of all, the Soviets had launched Gagarin on an orbital flight without the benefit of a world-wide tracking network. In other words, had Gagarin had a problem on his mission outside of the range of the Russian's tracking radars, they would have had little idea as to what had happened or where he might have landed. In addition, the fact that the astronaut had to eject from his spacecraft before landing indicated that their design was not fully developed. Representative Joseph E. Karth, a Democrat from Minnesota, gave a logical rational for why the Soviets beat the United States into space:

> *The United States and the Soviet Union have proceeded along two different lines of attack. The Soviets have pretty much rifled their program. . . as opposed to the United States shotgunning their effort. We have been interested in many programs and I think the Soviets have been interested primarily in putting a man into space.*
> — Representative Joseph E. Karth, April 1961.

While Gagarin was awarded the title of "Hero of the Soviet Union," it didn't keep him from losing his life in a two-seat MiG-15 training aircraft on March 7, 1968 at the age of 34. Gagarin, along with copilot Vladimir Seryogin, was killed in his aircraft shortly after takeoff from a military airfield in Russia. While the astronaut's death remains shrouded in mystery, Russian officials concluded that it was a result of pilot error. Seryogin was a senior test pilot and

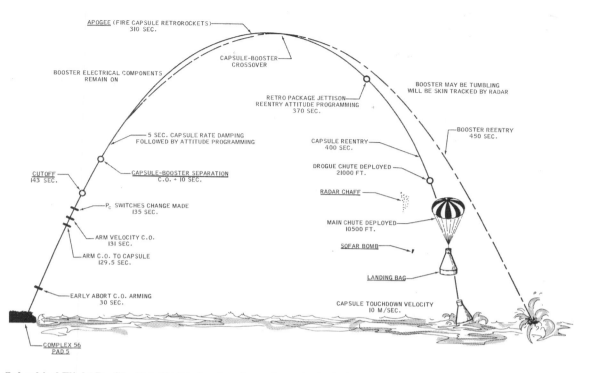

Suborbital Flight Profile: This NASA drawing shows the typical mission profile for a suborbital Mercury flight. Only two such manned missions were flown, though three were originally planned. The drawing accurately shows the radar chaff as well as the first SOFAR bomb which were both deployed before landing. NASA Drawing.

Heavy Lift: The 58-foot tall Redstone booster for Mercury-Redstone Mission No. 4 is lifted into place at launch pad 5 sometime before Grissom's Liberty Bell 7 flight. The gantry used to service the flight vehicle before launch was manufactured by a company with experience building oil production platforms. NASA Photo.

Umbilical: Pad Leader Guenter Wendt (left) checks out the spacecraft prior to launch. The umbilical cable for Liberty Bell 7 was attached through one large electrical connector mounted just under the optical periscope. When the cable separated during launch, a small spring loaded metal flap snapped shut, sealing the opening. NASA Photo.

Transfer Van: Grissom and Flight Surgeon Dr. William K. Douglas depart the NASA transfer van and make their way to the launch pad during one of numerous full-dress rehearsals conducted before the actual flight. The unit in Grissom's right hand helps keep the astronaut cool before he enters the spacecraft. NASA Photo.

Launch Pad Operations: Workers from NASA's Launch Operations Directorate surround Mercury-Redstone 4 at Pad 5, Cape Canaveral Air Force Station. The "cherry picker" positioned to the left of the capsule can be used in a emergency to remove the pilot just before liftoff. The modified crane is not relocated until T-55 seconds. NASA Photo.

Ready to Go: Grissom's Redstone booster and Mercury spacecraft stand fueled and filled with liquid oxygen early in the morning on July 19, 1961. The vapor surrounding the booster is boil-off from the oxidizer and the gantry has been moved back at T-270 to check radio frequencies. The particular effort was postponed due to bad weather in the launch area. NASA Photo.

On the Surface: The Squalus breaks the surface in 1939 after Navy divers attach several air filled pontoons to the crippled submarine. What followed was the grisly task of removing the bodies of the crew members who didn't make it. The salvage of the Squalus was a groundbreaking operation from the standpoint of crew rescue and the lifting of a large object from the 243 foot water depths. US Navy Photo.

Bathyscaph: The Trieste dangles on the end of massive lines while being lifted out of the water circa 1958 - 1959. Before the advent of syntactic foam, vehicles like the Trieste had to use large floatation tanks filled with aviation gasoline for positive buoyancy. The actual compartment inhabited by the two operators is the small spherical pressure vessel attached to the bottom of the vehicle. The two cylindrical structures on either side of the crew compartment are the vehicle's shot ballast system, used to dump steel shot which allows the submersible to surface. The Trieste could be made heavy by simply venting off gasoline. US Navy Photo.

The Deepest Dive: It was during one of the dives 200 miles southwest of Guam that the Trieste, operated by Navy lieutenant Don Walsh and Jacques Piccard, established the record for the deepest dive ever: 35,800 feet down in the Challenger Deep. The record stands to this day. The Swiss flag flying at the Trieste's stern signifies the participation of Piccard, a Swiss national. US Navy Photo.

USS Thresher Underway: It was the loss of the 593 boat that drove a crash program in undersea development when it became painfully clear that the United States lacked the ability to operate in the deep ocean, in this case, 8,500 feet down off New England. With the Thresher went 129 of her crew. US Navy Photo.

Underwater Grave: The upper rudder of the Thresher is visible in this 1964 photo taken by the Navy research ship USS Mizar. US Navy Photo.

A Grim Reminder: While the Trieste lacked the ability to recover any objects of significant size or weight from the Thresher, it did manage to pick up this section of brass piping from the submarine's debris field. The tube had been crushed by the tremendous pressures at depth, about 3,786 pounds per square inch. US Navy Photo.

A Pioneering Vehicle: The US Navy's CURV III Remotely Operated Vehicle was one of the first examples of the emerging ROV technology and instrumental in the recovery of a hydrogen bomb off the coast of Spain in 1966. Although primitive by today's standards, CURV earned a place in ROV history that was well deserved. US Navy Photo.

Missing Weapon: When a hydrogen bomb disappeared into the sea off Palomares, Spain in 1966, the US Navy pulled out all the stops to find the weapon, lost during a mid air collision between a B-52 bomber and a KC-135 tanker during a midair refueling operation. In addition to CURV, the Navy used the deep submersibles Alvin and Aluminaut, as well as similar vehicles supplied by Perry Submarine. In the end, while discovered by Alvin, the weapon was recovered after CURV became entangled in the bomb's parachute shroud lines. US Navy Photo.

Spook Ship: The Navy's Research Vessel Mizar (T-AGOR-11), originally a 1850-ton Eltanin class icebreaking cargo ship, was their premier deep water oceanographic tool during the mid 1960's and played major roles in numerous public and classified undersea operations. Among the Mizar's accomplishments were the finding of the nuclear submarines Thresher and Scorpion as well as the scuttled Liberty Ship Le Baron Russel Briggs in 16,000 feet of water in 1970. US Navy Photo.

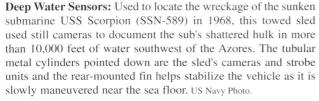

Deep Water Sensors: Used to locate the wreckage of the sunken submarine USS Scorpion (SSN-589) in 1968, this towed sled used still cameras to document the sub's shattered hulk in more than 10,000 feet of water southwest of the Azores. The tubular metal cylinders pointed down are the sled's cameras and strobe units and the rear-mounted fin helps stabilize the vehicle as it is slowly maneuvered near the sea floor. US Navy Photo.

decorated military hero who had flown more than 200 aircraft combat missions during World War II.

Shortly following Gagarin's historic orbital flight and taking into account the success of the MR-2 mission, NASA's Space Task Group was at the stage where they felt ready to attempt a manned space flight. What they did was train astronauts Shepard, Grissom, and Glenn for a potential suborbital flight and allow the remaining four pilots to focus on both ground support activities and training for orbital flight.

Although the nominal suborbital flight profile was designed around "old reliable," i.e., a Redstone booster, there was some earlier concern that all of the changes to the standard tactical Redstone rocket may have affected its performance, from the reliability standpoint. NASA selected the Redstone in the first place because it was an "off the shelf" design. However, after all of the modifications, it was now a missile nearly custom built for a manned Mercury mission. The military attacked the reliability problem in a different manner, by simply building and launching enough missiles to guarantee that any enemy target could be hit, no matter how many rockets failed. Obviously, NASA was not in a position to risk the loss of boosters (and astronauts) in a similar manner. As a result, STG engineers made a conscious effort to restrict the type and number of modifications made to their Redstone boosters; hopefully, this would reduce the possibility of aborted missions.

Fortunately for NASA, Yuri Gagarin's orbital flight had done much to silence the apprehension of the aeromedical community with respect to whether or not a man could survive a ride into space. Clearly, if Gagarin could do it, an American could as well. However, this didn't stop a newly elected President Kennedy from making inquiries regarding the potential risk of sending a NASA astronaut into space. From the political standpoint, the last thing Kennedy needed was to have an astronaut killed during the first American attempt at manned space flight. In fact, just before Alan Shepard's suborbital flight, Kennedy Press Secretary Pierre Salinger was calling Cocoa Beach, Florida, asking questions about the reliability of the Mercury escape system. Fortunately, Kennedy's concerns were satisfied and NASA continued with their preparations to get an American into space.

At 05:15 hours on May 5, 1961, former Navy pilot Alan Shepard made his way up the gantry to climb into his Mercury spacecraft, named Freedom 7; the number seven was added because his Mercury capsule was the 7th one built by McDonnell Aircraft Corporation. Surprisingly, the identity of the first American to make a flight into space was not known to the media or the general public until Shepard's mission was canceled two days earlier. This time though, there was no hiding who was destined to take the trip. Unlike Yuri Gagarin, there would be no secret regarding what happened to the first American into space as every second of the flight was being documented by both the domestic and international news media. In fact, over 350 news correspondents had registered for the opportunity to watch Alan Shepard either make history – or get vaporized on live television. As launch preparations continued, astronaut Wally Schirra circled the launch pad in his F-106 chase plane waiting to see what happened.

Shepard felt his heart rate quicken to 126 beats per minute as he watched the Mercury spacecraft's umbilical fall away. Feeling a deep rumbling sound underneath his back, he quickly punched the capsule's satellite clock, signifying the start of his flight. At first, things went smoothly for Shepard as he felt his body being pushed ever higher and faster, a sensation not unlike a catapult shot from a Navy aircraft carrier. However, as his craft punched through the Earth's ceiling, his helmet started vibrating like a jackhammer, to the point where he couldn't even read the dials on his control panel. Fortunately, the shaking subsided as Freedom 7 shot past the speed of sound on a trajectory off course by only one degree. So far, the flight was perfect. Millions of American's watched and listened to the events unfold, hopeful that America's new space hero would counter the Soviet space threat.

One hundred and forty-two seconds after ignition, Shepard's clone of a V-2 rocket shut down as it tossed the man farther upward at over 5,100 miles per hour. Though the astronaut looked, he was unable to see any smoke from his escape tower as it rocketed away from the speeding space capsule. Even though Shepard was anxious to stay ahead of his flight plan, he got irritated when he realized that he had forgotten to remove the medium gray filter from the capsule's periscope; he spent a few seconds fumbling with the filter knob until he noticed that his left wrist was banging into the abort handle. That killed it. Leave the damn filter alone and concentrate on the flight. Shepard the pilot was back in control of things as he raced to get ahead of his game and test the craft's manual attitude control system. A steel washer floating by his ear confirmed to the astronaut that he was in a weightless state. Surprisingly, he couldn't even hear the squirts of hydrogen peroxide gas as he manipulated the hand controller. As Shepard continued on his long parabolic arc over the Atlantic Ocean, he continued communicating with the Mercury Control Center at the Cape informing them of his sensations and progress.

Before he knew it, his space adventure was almost over as he felt the "comforting kick in the pants" when the three retro rockets fired in sequence, aiming Freedom 7 towards the US Navy recovery force awaiting his arrival. The drogue and main parachutes popped out right on time and only 45 minutes after landing within sight of the USS Lake Champlain, Shepard was listening to President Kennedy congratulate him on his successful flight. Shepard, along with fellow astronauts Slayton and Grissom, was flown to Grand Bahamas Island where he completed his post flight debriefing. In general, he experienced only minor problems on his mission such as a difficulty seeing his left wrist pressure gauge, a few nagging pressure points in his space suit, a little overheating after landing, and a tendency for the capsule to roll clockwise when in the manual control mode. What was interesting is that when Shepard switched to the Fly-By-Wire control mode, he neglected to shut off the capsule's manual control valve. As a result, when he maneuvered the spacecraft around in different directions, the Reaction Control System used valuable fuel from both the Automatic and Manual systems. This specific problem would come back to haunt NASA during Scott Carpenter's later orbital mission and it was this quirk in the capsule's design that caused Carpenter to use so much fuel during alignment for reentry. In other words, while it was possible for the astronaut to use both the automatic and manual Reaction Control Systems at the same time, the astronaut could easily end up with empty fuel tanks when it came time for reentry.

Overall, the citizens of the United States rejoiced at Shepard's success and the media ate it up. Soviet Premier Nikita Khrushchev was said to be less than pleased at the world wide attention NASA's first space mission received. After all, hadn't Gagarin orbited the Earth? What was all the fuss about a far less ambitious suborbital flight? The difference was that NASA's operations were done in full view of the world. Whatever happened, everyone was going to know about it and it's possible that this simple fact made their attempts appear more daring. How many chances were the Soviets willing to take if they knew that their failures would be transmitted all over the world? Didn't it take more guts to try something like this in public, as opposed to in secret? It did, and the world appreciated America's willingness to be judged in the international arena.

Now NASA and the guys at the Space Task Group had one under their belt. They had proved that they could send a man up there and get him back alive. If NASA, as an organization, was a fighter pilot, they had flown a combat mission, gotten shot at, and come back in one piece. Everyone had done their jobs and there had not been any major screw-ups. Now they had to do it again and prove that they didn't just get lucky. And Langley had just the man for the job: a guy from Mitchell, Indiana who didn't say a whole lot, but was worth listening to when he did. His name was Gus Grissom and he was rumored to be one hell of a pilot.

Chapter 4 – July 21, 1961
The Flight of Liberty Bell 7

If there are two or more ways to do something, and one of those ways can result in a catastrophe, then someone will do it.
— Edward A. Murphy

The spacecraft Grissom was scheduled to fly came off the assembly line at McDonnell Aircraft in May 1960. It was designated as McDonnell Capsule No. 11 and was more similar in design to the orbital version of Mercury spacecraft than the vehicle Alan Shepard flew earlier. Overall, the craft that Grissom would later call "Liberty Bell 7" incorporated numerous innovations, all of which would be tested in space for the first time.

Unlike Shepard's Freedom 7 capsule, which only had a couple of small 10 inch portholes for the astronaut to look through, Grissom's vehicle had a new and larger centerline window manufactured out of triple-paned Vycor glass from the Corning Glass Works. This window was as strong as any part of the capsule structure. There was also a redesigned fairing added to the capsule adapter clamp ring and a little more astronaut couch padding installed to help reduce the vibrations that Shepard had felt during his mission (he experienced blurred vision when trying to read gauges during liftoff due to the severe oscillations). In addition, Grissom would use a redesigned control panel as well as a new Rate Stabilization Control System; with it, the astronaut could control the speed of spacecraft movements by minute adjustments of the hand controller. There were also changes in Grissom's space suit in that his wrists were fitted with nylon-sealed ball bearing rings to allow for easier movement, a urine reservoir was added, and his helmet was fitted with new microphones designed to filter out more noise.

One final Capsule No. 11 innovation was an explosive exit hatch, designed and installed on the spacecraft at the request of the astronauts. The original procedure for exiting the spacecraft required the astronaut snake out through the parachute compartment, a difficult maneuver at sea in a bobbing capsule. The hatch, which had been thoroughly tested on the ground, would be flying in space for the first time.

Space Task Group engineers, while not concerned about the new hatch, were worried about the possibility that the spacecraft and Redstone booster might somehow crash into each other following separation. The basic suborbital flight profile was a long parabolic arc, where the Mercury capsule reached a maximum altitude of around 116 miles and landed about 300 statute miles downrange from the Cape, near the Bahamas. This allowed for a period of weightlessness lasting 4.4 minutes and a total flight time of 15½ minutes. The maximum speed was predicted to be a little over 5,000 mph and the capsule separated from the booster at an altitude of around 40 nautical miles. Capsule No. 11 was fitted with several Atlantic Research posigrade rockets created specifically to help in the separation process. What the engineers discovered from their calculations was that when these rockets were ignited inside of the spacecraft-booster adapter, they developed 78% greater thrust than when fired openly. This was called the "popgun" effect. They also estimated that the spacecraft would have a separation velocity of about 28 feet per second and be 4,000 feet away from the Redstone booster at retrofire. While it was thought that the booster would tumble into the ocean 16½ miles downrange from the capsule, they were still concerned enough about it to make minor changes in the retrofire sequencing.

Grissom had his first look at his new toy in January, 1961, about six months before his flight, during a visit to the McDonnell Aircraft plant. While there was much work remaining to do on the spacecraft, it wasn't until after Shepard's flight that Gus was able to really concentrate on his own mission. The Mercury capsule arrived at Hanger S, Cape Canaveral on March 7th and it was there that all Mercury spacecraft were prepared for flight.

> *Things moved so fast, however, that I did not get to do all of the things I wanted to do before the flight. I would have liked to practice more often on the ALFA trainer...*
>
> — Virgil I. Grissom

When Grissom's space vehicle was being manufactured, he made a point of attending the production meetings at the McDonnell plant in St. Louis, keeping tabs on the engineers and making sure that critical parts arrived on time from the subcontractors. He also figured it was useful to hang around during the building process as he ". . . thought it would be good for the engineers and workmen who were building my spacecraft to see the pilot who would have to fly it. . ." However, even with Grissom looking over their shoulders, the engineers still made mistakes:

The controls got snarled up somehow in the assembly at the plant, and the first time I tested them out the system yawed to the left when it should have gone to the right. The sort of thing should not happen even on automobiles. . . the attitude controls failed to center themselves again after a yawing maneuver and I had to compensate for this myself. . .
— Virgil I. Grissom

Even though his Mercury capsule was now at Hanger S at the Cape, Grissom still watched the spacecraft like a hawk to make sure that no one ripped anything apart that could potentially delay his flight. It wasn't that he didn't trust the engineers and technicians; on the contrary, they worked very long hours and he appreciated their dedication. But what he didn't want was to see some part of his new "ride" torn down to bits when it worked fine, but not to the level of perfection engineers always want. This is where Grissom's common sense approach to engineering really paid off. For example, there were minor problems with the capsule's electrical system in that an indicator light might not illuminate when the retro package was jettisoned. This happened on Shepard's flight and was thought to be caused by an overloaded electrical circuit. Though Bob Foster, one of McDonnell's engineers, had some fixes in mind, Grissom managed to nix the effort saying that he could override the problem just as Shepard had during his flight. Certainly, when a potentially dangerous problem cropped up, Grissom was usually for making things right. But the bottom line was that *he* was the pilot and used this fact to establish what level of risk was acceptable.

As Grissom and the engineers toiled at the Cape throughout the months of June and July, more and more minor problems threatened to delay the flight. First of all, it looked like the Redstone booster allocated for the mission might not be delivered in time, technicians were concerned about various components in Grissom's space suit, and then the pilot learned that the satellite clock, which keeps track of the mission's elapsed time, had rusted out. They had no replacement, until Sam Beddingfield, one of the NASA engineers, vowed that ". . . he would produce a clock by the next morning – even if he had to steal one." The next morning, they had a clock.

The days and weeks ticked off while Grissom engaged in minor skirmishes with the engineers and technicians to keep them from messing up *his* spacecraft. He was also concerned about the long hours the engineers were working to make the deadline. Grissom was all for hard work, but sometimes it got out of hand:

> *I remember John Shriefer... [he] was the McDonnell capsule chief from St. Louis, and one of the finest troops I ever met. I knew he had been knocking himself out getting the capsule ready and ironing out one problem after another. But I did not realize how worn out he was until one day we stopped for lunch and he was so exhausted he could not eat a bite. He just dipped his spoon into the soup. I took him right back to the Terrace Dune apartments... and asked his wife to keep him there and make him rest for 24 hours.*
>
> —Virgil I. Grissom

July 1st was a big day for Grissom as he got to see his spacecraft clamped to the top of the Redstone booster. From then on he spent most of his time at the launch pad either participating in or observing every test they made of the spacecraft.

In 1961, pad 5 of what was officially called the Vertical Launch Facility (VLF 5/6) at the Cape Canaveral Air Force station, was on the leading edge of space technology and it was from this location that both the Explorer I satellite and Alan Shepard had been launched. But in comparison to today's Space Shuttle launch *complex*, it bordered on "garage space flight." The actual launch pedestal for the Redstone, which was only about 300 feet from the firing control room, was nothing more than a small chest-high steel framework mounted over the top of a pyramid-shaped blast deflector in a small clearing surrounded by trees and bushes. All of the interconnecting cables ran from the pad to the firing control room through a cable trench covered with steel plate. They could even weigh the rocket using sensors in the utility room, which was directly underneath the pad and connected to this dug-out channel. The whole area was paved with refractory type concrete and there were four cinder block baffles at each corner of the launch area protecting massive water nozzles so an exploding rocket could be hosed down. Inside the firing control room were walls of consoles filled with toggle switches, vacuum tubes, transformers, electrical gauges, and circuit breakers, all situated around one small red button, which fired the rocket. The inside of this structure looked more like the lair of a mad scientist than a NASA blockhouse. Protecting the personnel inside was a roof made of eight feet of reinforced concrete, a heavy steel blast door and 4 inch thick quartz bullet-proof glass. The blockhouse at VLF 5/6 was designed to accommodate 55 people during a launch. However, during the Mercury-Redstone missions, a portable 20-ton capacity air conditioner was installed to help keep the blockhouse interior cool since as many as 75 people might be crammed into the area. For anyone who has been inside of the building, that's a lot of bodies.

In every way, the world from which Gus Grissom would soon be launched was far different than today. Computers were massive contraptions that filled whole rooms, like the IBM 7090 mainframes at the Goddard Space Flight Center. Calculators? There were slide rules. There was no such thing as satellite navigation because there were no navigational satellites. The aircraft carrier USS Randolph, a veteran of the Second World War and the prime recovery ship, fixed their position in the splashdown area using Loran A (a form of radio navigation) and sextants. Cable TV was something you used when you were in the middle of nowhere and had to put a quarter in a hotel room TV set. The satellite clock inside of his Mercury capsule was hand wound and the Earth path indicator consisted of a small plastic globe rotated by dozens of small brass gears. Grissom's Redstone booster was almost a direct copy of the German Army's 1945 V-2 rocket and used the same fuel; it didn't even have a gimbaled exhaust nozzle, instead relying on graphite vanes for guidance. In fact, the *escape tower* used on the later Apollo Command Module had more thrust. The Beatles were still some unknown British pop band in Liverpool and if you wanted money, you had to go to a bank during business hours instead of an ATM machine. People bought 33-1/3 speed vinyl High Fidelity records instead of CD's and had never heard of 8-track or cassette tapes. Television was black and white and the NBC television footage taken of Grissom's splashdown was kinescopes. There was no calling Houston and saying ". . . we have a problem," because the Johnson Space Center didn't exist.

However, for all of the things Grissom didn't have, unlike the doomed crew of the Space Shuttle Challenger, there was something he did have: a method of escape during almost every phase of his mission. If anything happened on the launch pad, he could blow the hatch and get away using the cherry picker positioned right next to his capsule (the cherry picker was a modified articulated crane fitted with a rescue basket on the end). Or, he could fire the escape rocket at any point during the flight (before it was jettisoned) and land safely using the spacecraft's parachutes. If after reentry, both of the parachutes failed, he could blow the hatch as he plummeted to Earth and bail out using his own personal parachute. He had *options*. However, no matter what happened, it would all be over in about 15 minutes or less.

A final test at the Cape had Grissom entering his craft in his space suit for a dress rehearsal to guarantee that all of the radio frequencies they were using were compatible. This was important because stray radio frequencies could accidentally ignite some of the craft's pyrotechnics. For the first time, the astronaut got to see what it was like to sit alone on top of a missile. However, it was not until they moved the steel gantry away from the Redstone that Grissom *really* felt what it was like:

> *There I was, perched up on top of that slender 70-foot booster with nothing to hold it upright and steady but a half-load of fuel in the tanks. I think a fairly strong wind could have pushed me over, otherwise... as the gantry moved away... I had quite a start. It looked as though the gantry was standing still and I was moving... I got on the radio and said, 'it looks like I'm falling.' Bill Anderson, the NASA doctor who was on duty in the blockhouse, told me later that Al Shepard had experienced exactly the same sensation.*
> — Virgil I. Grissom

One of the least important things Gus had to do was decide on a name for his spacecraft and as the pilot, it was his prerogative. He decided on "Liberty Bell," mostly because of the capsule's bell-shaped structure. John Glenn suggested that the number 7 should also be added to signify the unity of the seven Mercury astronauts. Shepard's Mercury capsule had used the number seven at the end of its name because it was McDonnell capsule number 7. Should Grissom call his capsule Liberty Bell 11? That would not do. McDonnell Aircraft Capsule number 11 became Liberty Bell 7. As a final gesture, a crack copied from a fifty-cent piece was added to the side of the spacecraft to complete the effect. Liberty Bell 7 was the one and only American spacecraft launched with a crack painted on the side of it.

There was to be one final hassle at a meeting just before launch. The engineers had discovered what *they* considered a minor problem with the oxygen system. Even though Grissom had developed a good way to solve the fault, as usual, the STG guys wanted to tear the system apart until it was . . . *perfect*. Grissom, knowing he was every bit the engineer they were, stood his ground and stared them down. As he later wrote, "I was against letting them tear the system apart and risk more delays trying to fix it. I guess I got rather impatient. It was a Friday evening. . . so 45 minutes before my plane was due to leave, I got up and said as firmly as I could, that I was satisfied that the capsule was in good flying shape. 'Please don't anybody fiddle with it over the weekend,' I said. Then I went home."

Even though he was supposed to be taking the weekend off at home, Grissom still drove to Langley to get some more time on the ALFA trainer, where he could practice manual control of the capsule. The astronaut returned to the Cape Sunday morning and continued a furious sequence of tests, all of them simulated abort missions. All that week he attended more meetings, ran more practice missions, and even managed to find time to do some fishing. However, there was still more frustration in store for Grissom:

> *We had our booster review with Dr. Jack Kuettner, one of the former German scientists at Huntsville...Apparently the bird had been behaving well in all of its tests...Then we had another meeting to decide if we were still agreed on all of the things we had already decided to agree on.*
> —Virgil I. Grissom

Twice before Grissom was actually launched, his mission into space was delayed. The first time, he was lucky in that the launch was scrubbed before they fueled the Redstone booster. That meant there was a delay of only 24 hours, as opposed to 48. On the next attempt, 4:15 AM on the morning of July 18th, 1961, Grissom was riding out to the launch pad sitting fully dressed in his space suit in a common vinyl "lazy boy" couch in the NASA van. Space flight in the early 1960's was filled with many traditions, much as it still is today. Someone stenciled a sign just inside the door of the transfer van saying, "Shepard and Grissom Express." Flight physician Bill Douglas handed Grissom a crossword puzzle book with a note inside from Sam Beddingfield telling him that, ". . . since the flight load has been reduced, we did not want you to get bored." Gus slipped his body into the tight fitting contour couch and John Glenn handed him a note reading, "Have a smooth apogee. . . See you at GBI." Grissom complained to Guenter Wendt, the McDonnell pad leader, about the fingerprints on the capsule's window and someone said that "they would install windshield wipers for the next shot." Unfortunately, after four hours of sitting flat on his back, Walter C. Williams canceled the flight reporting that, "the weather was not good enough for the flight they had planned." Now they had to purge the Redstone of its corrosive fuels, dry it out, and start the process all over again.

On July 21th, once again Grissom was awakened in the Florida darkness at 1:10 AM. Once again he had a light breakfast and felt the biomedical sensors being glued to his body. Once again he stuffed his short frame into the combination of neoprene bladders, laces, wires, and nylon called a space suit.

And once again he took the ride up the side of the gantry wondering if he was going to return alive. This time there would be no more delays.

> *MR-4 was launched at 0720 e.s.t., July 21, 1961, after holds totaling 1 hour and 20 minutes...*
> — Postlaunch Memorandum Report for Mercury-Redstone No. 4

Because of all the previous cancellations and holds, Grissom must have been somewhat tense as he crammed his space suit clad body into the tiny six-foot diameter cone-shaped capsule. Christopher Kraft, the head of flight operations for Project Mercury later remarked that, "Anyone getting on top of a rocket in 1961 would have to be scared if they understood the problems; and Grissom understood the problems."

> *Grissom's weight just before the flight was 150 pounds and 8 ounces, his pulse 68, his respiration rate 12 breaths per minute. His neck was 'normally flexible,' his thyroid gland was 'unremarkable.' The abdomen was 'soft, without tenderness. Eye, ear, nose and mouth examination was negative. Heart sounds were of normal quality, the rhythm regular'. . . no evidence of overt anxiety, that Grissom explained that he was aware of the dangers of the flight, but saw no gain in worrying about them. He felt somewhat tired, and was less concerned about anxiety than about being sufficiently alert to do a good job.*
> — Project Mercury Flight Physicians

The flight was delayed further when one of the 70 titanium bolts holding the capsule's hatch became misaligned during installation; it was removed so the mission could proceed on schedule; a memento Grissom kept as a souvenir.

> *The powered phase of the flight was normal. Booster cutoff occurred at 02:22 at which time the velocity was 6561 feet per second...*
> — Postlaunch Memorandum Report for Mercury-Redstone No. 4

Leaping off the steel launch pad, the Chrysler built Redstone rocket thundered into life and climbed into the Florida sky on the leading edge of a light contrail as Grissom radioed back, ". . . it's a nice ride up til now . . .," amidst the background of a slight rumbling sound.

There was a mild buffeting as the 80 foot long craft accelerated into the atmosphere through max-q, the term for maximum dynamic pressure. As the

Redstone cut off, Grissom was abruptly thrown against his shoulder straps, then shoved back into his seat as the posigrade rockets kicked Liberty Bell 7 away from the booster.

To the relief of the ground controllers and following capsule separation, Grissom, using his Collins transmitter-receiver, commented, ". . . I don't see a booster any place." For now, all was going according to plan.

Grissom, who was heavily strapped into what was termed the "forebody" area of the capsule, even had knee straps holding his legs in place, to keep his legs from flailing around during a potentially violent abort scenario, where the capsule would be jerked off the end of the booster by the powerful escape rocket.

While in place inside the capsule, Grissom was closely surrounded by numerous subsystems designed to both control the spacecraft and keep him alive. Within his reach was a control panel which allowed him to observe the functions of Liberty Bell 7 during the mission. In his right hand, he held a stubby control stick through which he could control the attitude of the vehicle. On his left was an abort handle, which in an emergency situation, could rocket the capsule up and away from the booster. Facing him slightly above his head was a large observation window used to see outside of the capsule so he could visually orient it in any desired attitude.

Also crammed into the small volume of Liberty Bell 7 were three oxygen bottles, seven miles of electrical wiring, two inert gas bottles for the attitude control system, a periscope for viewing the Earth, and two D.B. Milliken 16 mm film cameras for recording his reaction to the flight and the read-out of the gauges located on the control panel.

Grissom was separated from the deadly environment of space only by his thin suit and a 0.016" thick titanium pressure vessel and would soon be flying considerably faster than during any of his missions in military aircraft. The astronaut, however, was well acquainted with such encumbrances and dangers due to his long experience with testing jet aircraft.

Capsule separation occurred normally at 02:32 and after 5 seconds of rate damping, capsule turnaround was accomplished.

After being separated from the now tumbling Redstone booster, Grissom methodically began his program of maneuvering the capsule about all of its axes while he floated in zero-g. Not surprisingly, his attention was constantly being distracted by the incredible view outside the small picture window above his head; "... It's such a fascinating view out the window that you just can't help but look out that way."

He had seen the Earth as only two other men had. He was trying to concentrate on his instruments, but he just couldn't help looking out that damn window. He had even made a bet with astronaut Alan Shepard for a steak dinner that he would see a star during the flight.

Grissom thought he saw one, as he excitedly exclaimed, "... and I see a star." But it turned out to be only the planet Venus.

The brightness of the blue sky rapidly turned to the blackness of space. He was breathing hard as he went through the flight plan.

> *He performed well, and like Astronaut Shepard on the MR-3 flight, barely noticed that he was in a weightless condition... the pilot withstood the g forces during reentry without difficulty, making several voice communications during this period.*

Grissom struggled with the capsule's flight controls as he tried to test the vehicle in various attitudes. But to his dismay, he found the response "sluggish."

In his short-lived weightless state, he also noticed bits of debris, fragments of wiring, nuts and bolts, washers, and the like floating around him while he went through his flight plan, "... there's an awful lot of stuff floating around up here..." he reported.

Grissom strained to get the capsule set up for reentry by aligning the small nose of the capsule using his flight instruments. He was supposed to do it visually, but at the moment, he forgot as he instinctively used the spacecraft's attitude control indicators. It was pitch black inside of Liberty Bell 7, except for an occasional shaft of blinding sunlight which would momentarily illuminate his torso – threatening to blind him as he tried to see the capsule's control panel. Grissom pitched and yawed the capsule with the Bell Aircraft Reaction Control System as he was vaulted over the top of a long parabolic flight path. He started a roll maneuver, but canceled it at the last minute because he was getting behind in his flight plan.

The flight controllers on the ground became momentarily concerned as Grissom exclaimed, "I'm in... not very good shape here..." They thought he meant that he was having some sort of trouble in space. But in reality, he meant to say that he was unhappy with his flight attitude as he aligned Liberty Bell 7 for reentry.

Grissom was slightly misaligned when he fired the retro-rockets – too much to his right and too far up – an error that would make him land slightly farther downrange and to the north of where he was supposed to impact the Atlantic Ocean.

Grissom calmly ignited the three Thiokol retro-rockets and set the spacecraft up for a slow roll rate during his re-entry into the Earth's atmosphere. The rockets fired in sequence with a roar as Grissom called off the event, "There's one firing. . . there's two. . . all three retros have fired out. . ." As the tiny capsule plunged back into the Earth's thick atmosphere, Grissom kept it on track by nudging the control stick and firing small pulses of hot decomposed hydrogen peroxide gas out of Liberty Bell 7's attitude control thrusters. Liberty Bell 7 spun back towards the Earth, shock waves forming around its blunt beryllium heat shield creating a contrail miles in length.

Soon, a small red light began glowing on his control panel telling him to get ready for the massive forces of reentry – the .05 g light indicating 5% of gravitational force. The pressure came on rapidly as Liberty Bell 7 began decelerating. Grissom, though, kept up a clear and lucid conversation with ground controllers all throughout this period despite the fact that he was now being plastered into his seat by the force of over nine g's ". . . Roger, g's are building, we're up to six. . . there's nine. . . there's about ten. . ." In less than 30 seconds, Grissom's body weight had climbed to over 1,000 pounds. He spat out his words with great force as he felt his lungs and diaphragm squash during the reentry. The g forces Grissom was feeling were almost three times what the Space Shuttle astronauts now experience during their return to Earth.

As the astronaut plowed through the atmosphere towards the Atlantic Ocean, the carrier USS Randolph, the prime recovery ship, and several destroyers slowly maneuvered on station over 100 miles below about 302 miles off the Florida coast. Several FPS-16 precision tracking radars located on the east of Florida and in the Bahamas monitored Grissom's location.

The loud report of a sonic boom gave Navy recovery forces the signal they had been waiting for as the space capsule fell like a stone towards the ocean. At last, they were able to see the Mercury capsule sinking towards the ocean during the final few minutes of flight, observing Grissom's thin contrail as a nearly vertical line stretched between the ocean and blue sky.

Grissom waited for the capsule's drogue parachute to pop out. His silver gloved hand must have fingered the switch with anxious anticipation until with a thump, the chute popped out at 20,000 feet altitude, right on schedule. The six-foot diameter parachute quivered and stabilized the falling capsule as it continued falling towards the Atlantic Ocean at over 400 miles per hour. The next big event was the deployment of the 63-foot diameter main ring sail parachute. If the chute failed, Grissom would be forced to fire a mortar charge to open the second reserve chute.

However, at an altitude of about 10,500 feet, the main parachute billowed out, finally slowing Grissom down to a relatively sedate descent velocity of 32 feet per second while the pilot exclaimed, ". . . main chute is good. . . main chute is good."

Aboard the Randolph, Captain Harry E. Cook, Jr. strained his eyes to locate the capsule's ring sail parachute. Simultaneously, Marine Lieutenant James Lewis and copilot John Rinehart took off in one of four recovery helicopters with one objective in mind; recover the capsule and the astronaut inside it. There was about an 11 knot wind blowing with relatively calm seas and good visibility – excellent conditions for the recovery they had practiced so many times.

> We had rehearsed these procedures and activities in the bay off Langley AFB with the astronauts, and there was no reason to not be calm. That's what good training does. In addition, Gus had a low resonant voice, which was pleasant to hear. All the communication was of a type that one hears on any such voice transcript. Very calm and very professional.
> — Jim Lewis, pilot, Hunt Club 1.

Grissom, in the meantime, was still progressing through his flight plan verifying switch positions and inspecting a small hole in his parachute, "Ah, roger, you might make a note that there is one small hole in my chute. . . it's a triangular rip, I guess."

Shortly before landing, the heavy heat shield fell from the bottom of the black conical spacecraft with a thump as Grissom reported, ". . . 32 feet per second [descent velocity] and the landing bag is out green . . ." Liberty Bell 7's heat shield was designed to deploy from the capsule's bottom on the end of the impact landing skirt. The device would soften the craft's landing into the ocean and help keep the capsule upright in the seas.

Grissom, viewing the ocean surface through the optical periscope, described the landing of the capsule by commenting, ". . . I can see the water coming right on up!"

Landing occurred on schedule at 15:37. Astronaut Grissom asked the helicopter in the recovery area to stand by until he had recorded all switch positions.

Liberty Bell 7 splashed into the warm Atlantic waters and immediately fell over on its side, submerging the capsule's window in the process. All Grissom could see through the window was gurgling water, as the light wind dragged the spacecraft along in the seas. Grissom hurriedly jettisoned the main parachute by firing a small gas-power plunger, thus releasing the parachute. With the force of the parachute on the capsule gone, Grissom calmly waited for the capsule to right itself in the rolling seas. Finally, once he felt that Liberty Bell 7's small end was clear of the ocean, he flipped another switch firing a mortar charge which shoved out the now unneeded reserve parachute, also deploying a dye marker canister in the process. The water surrounding the bobbing capsule was soon filled with a vibrant green dye. A long spindly SARAH beacon antenna was also erected enabling the circling aircraft to take directional fixes on his location.

He was still talking to the recovery forces but was getting warm in the July heat, "I've unplugged my suit now so I'm getting a little warm . . ."

Observing Liberty Bell 7 splash down about six miles from his carrier, Captain Cook ordered his ship's speed up to 25 knots and headed directly for the floating spacecraft, which by then was being circled by Lewis' helicopter recovery force, designated "Hunt Club."

Unlike the astronaut portrayed in the movie "The Right Stuff," the real Gus Grissom had a calm voice and deliberate actions as he began recording the switch positions on his Mercury control panel and powering down his spacecraft. He was not in any hurry to get out of Liberty Bell 7 and was in fact telling the helicopters to wait while he finished up with his work for the day ". . . I'm just going to put the rest of this stuff on tape and then I'll be ready for you, in just about two more minutes I would say."

What happened next was an event that has never been explained. Grissom claimed that he was lying quietly in his couch. Others later said he was moving around inside the capsule and accidentally hit something. Still others said that he panicked, and for some reason had to get out of Liberty Bell 7. What is known is that he pulled a Randall survival knife off of the hatch and stuffed it into his survival kit, as a souvenir (this was later found underneath Grissom's astronaut couch). All during this time period, the two 16 mm film cameras inside of the capsule recorded his every move. Jim Lewis, in his UH-34D recovery helicopter designated Hunt Club 1, had begun his racetrack pattern approach to Liberty Bell 7 and was moving in for the hookup, but was still some distance away when he observed the hatch blow prematurely.

> *During the recovery operation the side hatch was seen to separate from the capsule followed soon after by Astronaut Grissom.*

The basic recovery plan called for the helicopter to hover near the top of the bobbing capsule and first cut off the long HF antenna sticking out of what was called the recovery aids section. This was done using shears at the end of a long pole called a shepherd's hook. Then, Rinehart would use the shepherd's hook to attach the helicopter's lift-line to a small dacron loop on the capsule top. After Lewis had lifted the bulk of the Liberty Bell-7 clear of the seas, Grissom would blow the hatch and be lifted up to the chopper with the sling. That was the way it was supposed to go. It didn't.

> *With the hatch gone the capsule took on water and immediately began to sink.*

Up to this point in the mission, it had been a "textbook" flight. Everything that was supposed to happen, did happen. But what followed next threw a wrench into the works. As copilot Rinehart told it, ". . . I saw beyond the capsule in the direction the wind was coming from, the hatch blown off the spacecraft. It flew about five feet flat and then turned and went skipping across the water." Grissom, who by his account was lying quietly in his couch, got out of the sinking capsule so fast that he didn't remember how he got out.

> *I was not worried about Gus being in the water. Because we had trained on these procedures at Langley AFB and STG and we knew the astronauts floated very well in their suits. . . At that point, we no longer had communication, so there was no way for any of us to know there was an open port in his suit. My plan at that point was to have my co-pilot cut the HF antenna (which was so long it would have been hit by the overhead rotors) with a device we had onboard just for that purpose, and try and snag the capsule before it sank. There was probably a minute or less from the time the hatch blew until it disappeared below the surface.*
> — Jim Lewis, pilot, Hunt Club 1

"I was just laying there (sic) minding my own business, and POW! – the hatch went." With the hatch gone, Grissom was faced with a life-threatening emergency; if he didn't get out of Liberty Bell 7, and fast, he would drown for sure. He chucked off his space helmet, ripped off the oxygen hose tying him to the vehicle and half floated and squeezed through the small hatch opening and struggled to get clear of the flooding capsule. He became tangled up in the Liberty Bell 7's dye marker canister, and fearful of being dragged down with the spacecraft, frantically pulled it clear of his harness. The Mercury capsule was shipping water, and fast.

> The helicopter hooked onto the capsule and attempted to lift it for a period of 3 to 4 minutes.

By now only about two feet of the top of Liberty Bell 7 was sticking above the surface as Lewis, in a dangerous maneuver, submerged the wheels of the chopper next to the flooding capsule. His co-pilot quickly chopped off the long antenna and showed the worth of all of his training by grabbing the recovery pole, turning on the chopper's movie cameras, and snagging the loop in record time. He had made the hook-up while the capsule was totally submerged; an amazing feat.

> I had to put the wheels in the water (the aircraft was not designed for this) after my co-pilot cut the antenna so he could reach the recovery bale on top of the capsule. By the time he had made the connection between the helicopter recovery line and the capsule recovery loop, the top of Liberty Bell 7 had disappeared below the surface. I began attempting to lift it out of the water at that point, although I knew that the combined weight of the capsule and water was more than the lifting capacity of the UH34D helicopter. My co-pilot began lowering the hoist for Gus at that point, according to procedure, so that we could bring him aboard the aircraft. I hoped that I might be able to drain enough water from the capsule and landing bag to be able to lift it and fly back to the carrier. However, there was considerable water below the hatch level that would never have drained, and even when some water drained from the landing bag, a wave would catch it, pull us down enough for water to flow back into the landing bag.
> — Jim Lewis, pilot, Hunt Club I

Grissom, who had forgotten to close an oxygen inlet valve in his space suit, was also sinking, and the worst part of it was that no one knew it.

Finally, after some strong swimming, Gus was able to reach the vicinity of Lewis' horse collar. But now Lewis had another problem, because he had gotten a "chip" warning light indicating that there were metal fragments in the helicopter engine's oil sump. Lewis desperately tried to move the chopper away from Grissom so if he lost the engine, the astronaut would not be killed by the crashing chopper. Lewis told Rinehart, "Bring the hoist back in, we've got a sick bird." Grissom, not knowing this or that Lewis had called in the backup helicopter to get him, could only see that the helicopter supposedly responsible for rescuing him, was pulling farther away. Grissom was drowning and the worst part was that no one seemed to realize he was in trouble.

He frantically waved to the other helicopters as waves of salt water cascaded over his head. He was beginning to swallow gulps of water as he tired from several minutes of strong swimming. While the astronaut was fighting the sensation, the sharp edge of panic began creeping into his body as he realized: *I'm going to die out here:*

> *I was getting lower and lower in the water all the time, and it was quite hard to stay afloat...There were three helicopters there. I guess there were four – I don't remember seeing but three. I was caught in the center of all three of them and couldn't get to any of them... I thought to myself, 'Well, you've gone through the whole flight and now you're going to sink right here in front of all these people.*
>
> —Virgil I. Grissom

Even though the winds in the splashdown area were only 11 knots and the seas running about two feet, being in the open ocean under those conditions is considerably different than a trip to the beach, especially when you're wearing a bulky space suit filling with water, heavy boots, and trying to stay afloat without the benefit of any sort of life preserver. For a while, Grissom was able to keep his head above water by swimming. But as he began to tire, he started swallowing more and more salt water, all the time being slammed by the rotor wash from the helicopter directly over his head. He started spending more time under the water as opposed above it and he finally realized that he was in the process of drowning and there was nothing anybody could do about it. Even with several helicopters circling around him, he was on his own. And that was a scary thought.

Finally, Grissom was saved by a second helicopter while Lewis tried to keep the Liberty Bell 7 and his helicopter out of the water. Although he didn't remember it later, Grissom was dragged for several feet across the surface before he was finally lifted clear of the seas. The first thing he did after being rescued was strap on a May West life preserver.

Lewis, in the meantime, was still trying to get the capsule clear of the water. But it was no use. He had the flooded spacecraft nearly drained many times, but simply could not keep it suspended above the seas long enough to drain the water out of the capsule's landing bag and then translate to forward flight.

> *I was pulling maximum power at this point (2800 RPM and 56.5 inches of manifold pressure). Shortly after I began this process, a chip detector warning light came on. This light indicated there were metal filings in the oil system. Our standard operating procedure for this event said that the engine would probably last about 5 minutes with metal being distributed throughout the engine before it failed... Water egress from a helicopter down in the water with rotors turning overhead is neither a risk free nor an easy task. I called the backup helicopter and told him to come in and pick up Gus and also said that I would drag the capsule clear of Gus so he could come in and make the pickup. Dragging it away was not that easy but we managed to get clear in a couple of minutes.*
>
> — Jim Lewis, Pilot, Hunt Club 1

Unlike the Gemini and Apollo capsules, Mercury's were fitted with an impact landing skirt. It was a cylindrical fabric construction which was deployed after the spacecraft was hanging on its main parachute. After softening the capsule's landing into the ocean, it functioned as a sea anchor and helped keep the vehicle upright during recovery. Unfortunately, it also held a lot of sea water – over four-tons worth. Despite Lewis' best efforts, that weight added to the ton of water contained inside the flooded Liberty Bell 7, made it physically impossible for him to lift the capsule clear of the seas, even under perfect conditions. Now he had a chip warning light, dropping oil pressure, a rapidly overheating engine... it was hopeless.

On board the Randolph, as soon as it became apparent that a real emergency was developing, Captain Cook advised Lewis that, "if it was getting dangerous... cut it loose and let it go."

> *...due to insufficient lifting capability for a water-filled capsule and an engine warning light, recovery of the capsule was abandoned.*

Lewis continued his efforts until he observed engine oil pressure dropping and cylinder head temperature rising. At that point, having heard the radio transmissions from the backup helicopter Hunt Club 2 that Gus was safely aboard and four minutes after successfully attaching the capsule to his helicopter sling, he radioed the carrier, declared an in-flight emergency and activated the switch to open the hook attached to the capsule's Dacron recovery loop, thereby releasing the capsule. Thus freed, Liberty Bell 7 splashed into the water on its side, slowly uprighted itself, and began a long three-mile descent to the sea floor. Lewis' chopper, free of the overload, lunged upward and forward across the ocean. He landed on the carrier minutes before a dripping Astronaut Grissom landed in one of the other helicopters.

> *The backup helicopter co-pilot lowered his hoist, and when Gus saw that, he figured out what was happening and began making his way toward it. I was pointed into the wind at this point and the backup helicopter was behind me also pointing in to the wind [this gives any aircraft added lift], so I could no longer see Gus... [he] was aboard that aircraft and on his way back to the Randolph in less than 4 minutes after the hatch blew on Liberty Bell 7, so you can see our contingency procedures worked perfectly. This is exactly why we had a backup helicopter close at hand. I had determined to continue trying to lift the capsule until my engine gave signs of quitting. After close to 5 minutes of pulling max power, the cylinder head temperature began to rise and the engine oil pressure began to drop and I made the decision to release the capsule so that I could set the helicopter down "normally" in the water if the engine died. I declared an emergency at that point and proceeded back toward the Randolph and was able to land aboard the carrier.*
>
> — Jim Lewis, Pilot, Hunt Club 1

In spite of the warning light seen by Lewis while he tried to save Liberty Bell 7, nothing was ever found wrong with the helicopter's engine. Grissom, who was in mild shock from his near drowning, was given a round of physicals aboard the carrier and pronounced to be in "good shape." However, the astronaut's post flight medical statistics are a better indication of his condition after nearly drowning in the Atlantic Ocean:

> *...The findings disclosed vital signs of rectal temperature of 100.4° F; pulse rate from 160 initially to 104 (supine at end of examination); blood pressure of 120/85 LA sitting, 110/88 standing, and 118/82 supine; weight of 147.2 pounds (the astronaut lost 3½ lbs), and respiratory rate of 28. On general inspection, the astronaut appeared tired and was breathing rapidly; his skin*

was warm and moist. Eye, ear, nose, and throat examination revealed slight edema of the mucosa of the left nasal cavity and no other abnormalities.
— Dr. Robert C. Lanning, USN, USS Randolph

Both Lewis and Rinehart later received letters of commendation for their efforts to save Grissom and Liberty Bell-7. In reality, it was a miracle that they managed to attach a line to the sinking spacecraft at all, never mind complete the recovery. Also, this was not the first time that Jim Lewis had helped save a NASA spacecraft that was in trouble, having been the recovery pilot for the Little Joe 5A flight off Wallops Island earlier in 1961. After hearing over the communications net that the Mercury capsule's parachutes had deployed prematurely (at maximum dynamic pressure), Lewis was able to autorotate his helicopter down along with the spacecraft and photograph the badly torn parachutes.

Astronaut Grissom's first appearance on the carrier and his initial statements indicated that the period when he was in the water was the more difficult portion of the mission.

Why Liberty Bell 7's hatch jettisoned prematurely was never determined. Grissom was adamant that he did not trigger the explosive hatch, but was never totally vindicated. The Postlaunch document suggested numerous possibilities, such as the "release of the external hatch-actuation lanyard from its stowed position, omission of the O-ring seal on the detonator plunger, chemical action of sea water on the explosive, galvanic voltages caused by action of the sea water and dissimilar metals, static electricity discharge between the helicopter and the capsule, as well as a differential pressure across the plunger." The report soberly concluded that "the hatch needs further study and test to prevent future premature opening."

Grissom later commented, "It remains a mystery how that hatch blew and I am afraid it always will. It was just one of those things." In fact, the spacecraft was probably still free-falling towards the sea floor when the Navy task force departed the area, being marked only by a smoldering smoke bomb and a trail of fluorescent green dye.

What happened to the spacecraft when it sank? The capsule probably reached its terminal descent velocity at a depth of less than 50 feet in the range of 3 to 4 miles per hour. At 1,000 feet, Liberty Bell 7 was in near darkness and still free-falling through the dense sea water. Somewhere below 5,000 feet the capsule's two fiberglass nitrogen bottles imploded with muffled bangs, bouncing shards of fragments off the back of Grissom's control panel. At about the same time, some of Liberty Bell 7's interior stainless steel piping was squashed by tons of water pressure. Sixty to 90 minutes after it vanished beneath the rolling waves, Liberty Bell 7, still intact, landed gently on a high-point of a rugged bottom called the Blake Basin.

As the years passed, the beryllium heat shield corroded away and small chunks of beryllium oxide were slowly separated from the structure and eased down the slight slope like tumbleweeds. Over time, assorted primitive marine organisms attached themselves to the capsule's outer shingled skin. After almost four decades on the bottom, the small end of the capsule became festooned with formations of white corrosion from galvanic action between dissimilar metals as did the aluminum control panel. Liberty Bell 7 was slowly covered with a light dusting of decomposed marine animals, with the writing "United States" and the painted crack still visible on its conical section.

For over 40 years, the question of why Liberty Bell 7 sank has continued to be debated by historians and space enthusiasts. Most of these debates center around Grissom and the controversial explosive hatch.

The explosive hatch fitted to Grissom's capsule was introduced into the design of the Mercury spacecraft as a result of complaints by the astronauts that there was no satisfactory way to exit the vehicle in an emergency. The original design of the spacecraft required that the pilot exit via the small end of the capsule, through the area that held the main and reserve parachutes. However, the procedure called for the astronaut to remove a section of the control panel, pull out a small pressure bulkhead, detach some electrical connectors, and then push out a fiberglass liner that had formerly held the two parachutes. It was altogether a very time consuming operation that was deemed unacceptable. This forward hatch was used only once during Project Mercury after the introduction of the explosive hatch, by Scott Carpenter following his mission in Aurora 7. Thus, the explosive hatch was created as part of the Mercury design.

Despite what many people think, the Mercury hatch was not held in place by "explosive bolts." Rather, the hatch, weighing about 23 pounds, was fastened to the side of the capsule by 70 titanium bolts, each of which had been modified (and purposely weakened) by the drilling of a small hole through the bolt shafts. The hatch was sealed by a magnesium gasket and an inlaid rubber seal. In close proximity to the bolts, a length of Mild Detonating Fuse (MDF), a rope-like explosive, was fitted into a channel that surrounded the perimeter of the hatch. The two ends of the fuse terminated at the hatch plunger mechanism, which was mounted on the inside of the door. Following actuation of the hatch plunger, the MDF would fire, causing the hatch bolts to fail in tension.

The hatch plunger was a relatively simple device that used two firing pins and percussion caps to detonate the fuse described above. Operation of the plunger took place in two stages: Initially, the astronaut had to remove a small screw cap that covered the plunger (which looked like a large push-button). At this point, the mechanism was "safed" by a safety pin and required 40 pounds of force to fire. The astronaut could then either fire the hatch, or remove the safety pin and depress the plunger with only four pounds of pressure. Removal of the cap and safety pin was tantamount to the process of removing the safety from a "cocked" revolver. However, the plunger mechanism could also be fired from outside the capsule without any action by the astronaut. This aspect of the design made it possible for rescuers to blow the hatch and save an incapacitated pilot. The rescue personnel would have to remove a small screw from the outside of the hatch and pull out a wire lanyard with sufficient force to break the interior safety pin (and fire the plunger). The problem was that the lanyard was only 42 inches long, and the hatch could easily travel 20 feet from the side of the spacecraft. As a result, NASA post recovery procedures described elaborate rope and pulley arrangements enabling shipboard personnel to safely fire the hatch from the outside. To mitigate danger from a rapidly traveling projectile-like hatch, it was also held in place from the inside by two short lengths of soft iron coils, which would normally limit the distance the door would fly to about a foot.

On Grissom's mission, the flight plan called for the astronaut to install these coils while the spacecraft was floating on its main parachute. Unfortunately, Grissom was unable to complete the task with his space suit gloves on. Thus, when the hatch detonated, it sailed off the side of Liberty Bell-7 and sank.

During Grissom's NASA postflight debriefing, he admitted that he had already removed the plunger cap and safety pin and was simply waiting for the recovery helicopter to call and tell him that they had hooked onto the capsule.

> "I was just waiting for their call when all at once, the hatch went. I had the cap off and the safety pin out, but I don't *think* I hit the button. The

capsule was rocking around a little, but there weren't any loose items in the capsule, so I don't see how I could have hit it, but *possibly I did.*"

However, later in the debriefing, Grissom discounted even the possibility that he could have accidentally hit the plunger. This fact was supported in part by astronauts John Glenn and Wally Schirra, who both used the explosive hatch on their Mercury orbital flights. Because of the recoil of the hatch plunger (from the explosive gases), they had suffered minor injuries to their hands during the firing of the hatch. Grissom was given a through physical examination after his flight and although he had sustained no such injury, parts of his body were well protected by certain elements of his space suit.

Most theories as to how the hatch could have fired on its own have to do with the possible release of the external actuation lanyard from the side of Liberty Bell 7 or the existence of a vacuum in the hatch plunger mechanism. If the lanyard handle had become dislodged during the flight, then the end (and the small handle) would have been dangling down near the capsule's landing bag straps while Liberty Bell 7 floated in the seas. With Grissom pulling out the safety pin, it is certainly possible that the handle could have been snagged by one of the straps, causing the hatch to fire. In another theory, the cable used to make up the external lanyard was sealed into the outside of the plunger mechanism by potting compound. If the sealant had leaked slightly in space, then a vacuum could have been created inside the plunger which may have made it move on its own after Grissom removed the safety pin. The only problem with this idea is that Grissom had indicated that he had pulled the safety pin some time before the hatch blew. Logic would seem to dictate that if a vacuum existed inside the plunger mechanism, the hatch would have detonated immediately following the removal of the pin.

If the capsule's external lanyard was hanging down the exterior of the capsule, NASA engineers should have been able to see it by examining the numerous photographs taken while Liberty Bell 7 floated in the seas. If they had done that and found nothing, why was it still a possibility for causing the accident? It may be that exterior photos of the explosive hatch, prior to jettisoning, simply do not exist.

Taking into account Grissom's high heart rate during the mission, writer Tom Wolfe ignored reality and declared that Grissom, a veteran of 100 F-86 combat missions in Korea, one of the most experienced test pilots in the United

States, and a man having over 3,500 hours of flight time in high performance jets, had "panicked" and blown the hatch prematurely. Wolfe based this absurd theory on the fact that Grissom's heart rate was higher than the other astronauts during their missions. However, during his indoctrination into Project Mercury, it had once gone up to 200 beats per minute on the "treadmill" test. As a result, the fact that he had a history of an abnormally high heart rate tended to downplay its effect on his performance and Grissom's deep voice sounded calm and professional on the mission tapes. Was there another possibility for what happened?

One of Grissom's debriefings was done at Grand Bahamas Island, shortly after his space mission, and it was here that he was met by Robert F. Thompson who was in charge of Project Mercury recovery operations:

> *So we fly down to Grand Bahamas, and about the time I get there, they bring Gus in... and bring the two pilots in the chopper. There was a barracks there that we were going to do some debriefing in. I told Gus to go into the little private room in the front of the barracks there, and I talked to the two helicopter people, got their debriefing pretty quickly, and then went in and sat down and talked to Gus, just the two of us in the room...*
>
> *So I talked to Gus about what went on. Well, after about five minutes of talking to Gus and the little bit of conversation I had with the helicopter pilots, I was pretty sure what the problem was. As far as I'm concerned, the problem was Gus got out of sequence.*
> — Robert F. Thompson, Johnson Space Center Oral History Project
> August 29, 2000

When Thompson says "out of sequence," he's suggesting that Grissom armed the hatch before that step in the post landing procedure was supposed to be completed. However, from Grissom's perspective, it's unclear exactly how established the procedures were for the astronaut after splashdown. In fact, this was one of the problems cited by the astronaut during his MR-4 postflight debriefing at Cape Canaveral two days after his flight:

> *I had trained for everything up to the time the main chute deployed. From there on, I hadn't really had any training. I didn't have it clear in my mind what I would do. It wasn't clear in my mind what I should have been doing on the water although I went ahead and did the proper things, like taking out the safety pin.*
> —Virgil I. Grissom, Astronaut Debriefing at Cape Canaveral, Florida
> July 23, 1961

In other words, while Grissom did not do anything wrong, it's a matter of when he did what. Looking back on the incident 40 years later, it's convenient to speculate that Grissom should not have even touched the hatch plunger mechanism until the recovery helicopter had latched onto the floating capsule. Again, though, it's not certain that these procedures were sufficiently established for the astronaut to know when to complete each task of the post flight activities:

> Well, he wanted to do such a good job, that while he was waiting for Hunt Club to get everything ready, he says, I'll just take the cap off, and I'll pull the pin, but I won't push the plunger. But now he's in a bobbing capsule, all kinds of stuff in there, and it's right here. Now, did he push the plunger? Of course not, you know, but it's kind of like you had a gun with two safeties on it. You took the two safeties off and you put it up and you pointed it at somebody, but you didn't pull the trigger, right? So he merely got out of sequence, trying to do such a good job... Did Gus push the plunger? No, he didn't push the plunger. Did he get out of sequence? Yes. He told people that he got all ready, and he just shouldn't have done it until he was told to do it. It's just that simple. But there was no point in making a federal case, and we went on about our business.
> — Robert F. Thompson, Johnson Space Center Oral History Project
> August 29, 2000

In the final analysis, it's not unreasonable to conclude that Grissom might have hit the button with some part of his thick, bulky space suit without feeling the recoil of the plunger. The detonator required only four pounds of force and there were a few parts of his suit that could have withstood the recoil of the plunger without injuring Grissom's body. He heard the hatch go, but with the dynamics of the situation (i.e., the capsule rolling in the seas), probably didn't even realize that he had caused it. In other words, he didn't lie about it; he simply didn't know what really happened.

During the same debriefing at Cape Canaveral, Grissom also had some thoughts about how to keep future astronauts from potentially drowning after their space missions:

> You've only got two men on board a chopper, but you're sending three, four, and five choppers out. One of those choppers ought to leave a photographer off and carry a guy to be put in the water with a wet suit and flippers to help you.
> — Virgil I. Grissom, Astronaut Debriefing at Cape Canaveral, Florida
> July 23, 1961

Following the end of Grissom's sub-orbital flight, NASA solved the hatch problem with a procedural change. From that point on, the astronaut would not remove the plunger safety cap or pin until the recovery helicopter had hooked onto the spacecraft. Fortunately, NASA decided on the above fix instead of Alan Shepard's suggestion that the hatch be held securely to the side of the capsule by an interior cable and turnbuckle arrangement suspended across the astronaut's chest.

> *Gus and I flew on an S2F from the Randolph to Grand Bahamas Island (GBI) for the debriefing and then I flew back to the carrier. Gus understood all that had happened and appreciated the efforts that were made to bring it back. While we all would have preferred to have the spacecraft, what resulted, given the circumstances, represented excellent results. The mission could certainly be considered a success, because, given the fact that problems developed, all of our contingency procedures worked and we brought Gus back. The mission objectives to test the reaction control system, piloting techniques, and "to prove that Freedom 7 was no fluke," were all successfully achieved.*
>
> — Jim Lewis, Pilot, Hunt Club I

As for Grissom, no one ever publicly accused him of anything, mostly because the astronauts had become such public figures by that time that NASA would have been foolish to admit that *any* of their boys had done anything wrong. In fact, as history eventually showed, NASA had the utmost confidence in Gus Grissom by selecting him to fly the first Gemini and Apollo missions. While Grissom was very sensitive about the loss of Liberty Bell 7, he wasn't above joking about the incident. When asked to name his Gemini spacecraft, he initially came up with the "Titanic;" *that* didn't go over very well with NASA. Grissom finally settled on the (unsinkable) "Molly Brown."

NASA had an ambitious timetable for putting a man on the Moon and returning him safely to Earth; they no longer cared about Liberty Bell 7, and it was largely forgotten. Space historians declared it as gone forever and most underwater salvage experts doubted such a small object would ever be found. Many didn't even think there would be enough left of Liberty Bell 7 to make it worth recovering anyway. Still others said that it was probably blown to bits as it sank by an on board explosive device. Out of the 20 Mercury spacecraft built by McDonnell Aircraft, six were flown on manned missions. Of these six, five were on public display*:* Freedom 7, Friendship 7, Aurora 7, Sigma 7, and Faith 7. One was still missing.

Chapter 5 – Underwater Vehicles
Pushing the Underwater Envelope

No man will be a sailor who has contrivance enough to get himself into a jail; for being in a ship is being in a jail, with the chance of being drowned.
— Samuel Johnson

When Liberty Bell 7 sank, there was scant underwater technology available to attempt a recovery, even if the sunken capsule could be found; a very doubtful proposition in 1961. In fact, immediately after the capsule's loss, Martin Byrnes, a representative of the Space Task Group on board the USS Randolph, suggested that a "marker be placed at the point so that the capsule might be recovered later." Rear Admiral J.E. Clark soberly advised Byrnes that the water depth in the area was about 2,800 fathoms. In other words, forget it.

The only submersible vehicle in the world capable of reaching such depths at the time was the US Navy's Trieste I, which set the record for the world's deepest dive in 1960 (a record that stands to this day): 35,800 feet underwater in the Challenger Deep about 200 miles southwest of Guam. However, even though the Trieste was put on alert after the Mercury spacecraft's loss, any salvage plan was eventually abandoned:

> *One of the ironies of the Trieste program is that the Navy had a manned submersible that could reach any depth in the world ocean but no way to get it there... My refusal to use Trieste [on Liberty Bell 7] was not very welcome by higher authorities. But I put it to them simply, in my view the bathyscaph could not make the long voyage from the nearest Navy base... There was a high probability that we would lose it too. And it was just not the long tow, but after each dive there was quite a bit of servicing required. Battery recharging, adding more gas to the float, loading several hundred pounds of steel shot, etc. The logistics were not trivial.*
> — Don Walsh, Program Manager, Trieste I

Simply diving on the capsule was only part of the problem in 1961. The spacecraft would have to be located, this in an era before the advent of side-scan sonar. Even if Liberty Bell 7 could have been found, how would they have raised it? At the time, the US Navy had none of the high-strength kevlar lift lines used in deep water salvage operations today because they simply didn't exist. When considering whether or not Liberty Bell 7 could have been found and raised in

1961, it's a matter of balancing actual capabilities against operational feasibility. Could Liberty Bell 7 have been found in 1961? Possibly. If the Trieste made enough dives in the area they may have stumbled across the spacecraft. Would it have been practical? No. In the end, it didn't matter as all of the critical flight data was radioed to the ground anyway during the flight. While the capsule's loss was an embarrassment to NASA, they had no pressing need to fund a costly search and recovery effort.

While underwater search and recovery operations have been conducted since the beginnings of recorded history, it has only been through the application of advanced technology that salvors have been able to locate and retrieve objects from the ocean floor with any real success. However, before the advent of sophisticated undersea equipment, the sea, in particular the deep ocean, was considered a mysterious world, inhabited by sea monsters and other frightening creatures. The ocean was a place that swallowed up things such as whole fleets of ships, along with their doomed crews. Even today, the fact that massive man-made objects can disappear into the dark, cold abyss, never to be seen again, is a terrifying thought; one that perpetuates rumors of Bermuda Triangles, windows into other dimensions, powerful magnetic fields, and long-lost cities. Unexplored areas on early sailing charts were shown as being inhabited by sea serpents with huge jaws, ready to crush any ship that dared venture forth. Even the explorer Magellan was not immune to a misconception of the abyss. In 1521 he once tried to measure the depth of the Pacific Ocean by splicing together several lines and lowering a cannonball more than 400 fathoms down (2,400 feet). After failing to detect the bottom, he declared "that the Pacific was immeasurably deep." Surprisingly, the preconceived notions of the peril of the deep are not much different from the reality, as the danger of the environment makes it nothing to trifle with. The deep ocean is easily the most deadly location on Earth, in comparison to Mount Everest, the Sahara Desert, the jungles of Africa, and Antarctica. In no other place can you be killed as quick as in the abyss. As a result, the sea remains a daunting place, still harboring the remnants of aviators Amelia Earhart and Amy Johnson, big band leader Glenn Miller, balloonist Tom Gatch, the Revolutionary Warship Bonhomme Richard, Flight 19, and the Heavy Cruiser USS Indianapolis, to name a few.

Most diving operations done in the early 1900's were accomplished by divers using what is commonly referred to as "hard-hat gear," where men wore a heavy canvas dry-suit, breast plate, and copper diving helmet. The only breathing gas available at the time was air, which seriously limited the depths that could be safely accessed. This was due to the effects of nitrogen narcosis, the so-called "rapture of the deep" and oxygen poisoning, which can cause a diver to go into convulsions at high partial pressures of oxygen. Up until 1912, US Navy divers rarely went deeper than 60 feet, mostly due to limitations in the existing diving tables. However, that year Chief Gunner George D. Stillson established a program to test Professor J. S. Haldane's (a renown physiologist) decompression tables, developed earlier under the auspices of the British Admiralty. Three years of experimental diving operations in Long Island Sound eventually enabled Navy divers to go as deep as 274 feet under water. It was a good thing, because the submarine USS F-4 was lost near Hawaii about six months later and without Stillson's earlier work, the Navy's salvage divers would never have been able to work at the 304 foot depths. Later advances in the creation of an artificial breathing medium (i.e., mixed gases, such as combinations of helium and oxygen) enabled salvors to reach several hundred feet farther into the depths of the ocean. In 1972 and 1975, US Navy divers extended the open ocean record for depth to 1,010 and 1,148 feet, respectively. However, in 1977, this record was pushed even farther underwater when a team of French divers broke the 1977 record by surviving operations at 1,643 feet. Even though a team of Comex divers later managed to break the 2,000 foot mark during a simulated dive, basic human physiology dictates that humans can only go so deep. To truly operate in the deep ocean requires a submersible vehicle.

The development of functional submarines in the late 1800's and early 1900's was mostly limited to the work of John Holland and Simon Lake, with the former primarily interested in the submarine as a weapon and the latter initially concentrating on underwater salvage. Holland later sold the U.S. Navy its first commissioned submarine, the USS *Holland* (SS-1), while Lake, a competitor of Holland's, elected to concentrate on the European market (Lake also sold submarines to the US Navy in later years). While both Holland and Lake contributed to early submarine development, Lake's unique technical innovations, such as ballast tanks, diver lock-out compartments, periscopes, bottom deployed submarine wheels (like those on the US Navy's NR-1 research submersible) and even-keel hydroplanes, did much to establish the marine engineer as the "father of the modern submarine."

In addition to the work being done by Lake and Holland, starting in June of 1930, William Beebe, a scientist at the New York Zoological Society, pioneered the use of a submersible vehicle called the bathysphere. His deep diving vehicle was configured as a heavy steel sphere (only 4 ½ feet in diameter on the inside) fitted with glass view ports and lights. The bathysphere, which was fitted with a 400 lb. hatch, was lowered into the sea using a long steel cable and the 14 inch diameter hatch opening made it a tight fit for any normal sized man. However, in an era where divers were plodding along in waters a few hundred feet deep, Beebe, along with engineer Otis Barton, were making dives to more than *half a mile*. Unfortunately, if anything went wrong with Beebe's bathysphere, at that point in time and in those water depths, there was absolutely no hope of rescue. Beebe discovered many new species of dragon fish during his numerous dives in the device and later described his findings in the book, "Half Mile Down." Beebe was no coward, to say the least.

Deep diving developed more fully as a technology following the sinking of the USS *Thresher*, a nuclear submarine lost off the coast of New England during sea trials in 1962 in over 8,000 feet of water. The result of this one incident was essentially a crash program initiated by the United States Navy to establish greater access to the deep ocean. Fueled by government funds, a number of U.S. aerospace corporations developed manned submersible vehicles for science and military related applications. However, manned submersibles had their limitations and it wasn't long before the government and private industry started experimenting with remote technology.

While Beebe's bathysphere fits into its own unique category, there are really only about three basic ways to operate in the deep ocean: using either a Deep Submersible Vehicle (a DSV, which is an offshoot of early submarines), an Autonomous Underwater Vehicle (an AUV, something like an intelligent torpedo fitted with sensors), or a Remotely Operated Vehicle (an ROV, the unmanned or teleoperated version of the DSV).

Deep Submersible Vehicles are small submarines designed to operate in the ocean abyss, sometimes in waters as deep as 20,000 feet. They typically carry from two to three people, are controlled using electric propulsion units, can weight as much as 20 tons in air, and are used mostly for science related operations. The deepest diving DSV operating today is the Shinkai 6500, a research submersible operated by the Japan Marine Science and Technology Center (JAMSTEC); as it's designation suggests, the Shinkai can operate in waters 6,500 meters deep (a little over 21,000 feet). The famed manned submersible, Alvin, also fits into this category of undersea equipment.

Autonomous Underwater Vehicles are a fairly new addition to subsea equipment and are somewhat similar to Remotely Operated Vehicles. The big difference is that AUVs operate without any cable linking them to the surface, i.e., untethered. In general, AUVs are untethered robotic underwater vehicles which can be either programmed to perform specific tasks underwater or even commanded from the surface in real time. Most AUVs have the exterior appearance of a very long (and expensive) torpedo. Since there is no direct electrical or optical fiber link to the surface, real time video is impossible. However, some versions of these vehicles do have the ability to do "data dumps" of electronic still photographs or side scan sonar data for evaluation while the vehicle is still submerged. In many ways, AUVs represent the future of remote underwater technology and they are already starting to be used for subsea tasks done previously with towed sonars and ROVs, such as bottom surveys.

Remotely Operated Vehicles are teleoperated versions of DSVs. As the term "teleoperated" suggests, these vehicles are unmanned submersible vehicles which are controlled in real-time by surface operators. They are linked to the surface via an umbilical cable which transmits video and telemetry data to the surface and supplies electrical power to the underwater vehicle. These days, ROVs can range in size and complexity from a simple 50 lb. "flying eyeball" to massive tracked vehicles weighing over 30 metric tons. ROVs, depending upon their intended purpose, can also be fitted with hydraulic manipulators, high-resolution imaging systems, water jet burial tools (for burying submarine telephone cables), as well as a variety of specialized tools needed to interface with subsea equipment. The vehicles are normally positively buoyant and maneuvered on the sea floor using either electric or hydraulic propulsion units.

The Remotely Operated Vehicles (ROVs) we used to find and recover Liberty Bell 7 did not appear out of nowhere, nor did the capability to do any sort of manipulative work underwater using remote technology. The computer, electronic, hydraulic, and mechanical systems so critical to our work in the Blake Basin evolved from their modest beginnings in the 1960s over almost four decades, where it is now almost commonplace to do remote work underwater in waters as deep as 16,000 feet.

The creation of the modern-day ROV was directly related to the inherent limitations of both divers and manned submersibles as well as advances in data transmission, computers, video cameras, and electrical and hydraulic control systems. However, the real driving force behind ROV development was that the operation of any manned submersible vehicle (except for those that are nuclear powered) is limited with respect to how long it can stay submerged and the fact that for it to function, it must be directly controlled by human operators. The endurance limitation is directly influenced by life support restrictions (i.e., the available supply of oxygen and a suitable carbon dioxide absorbent), indirectly related to the number of batteries and/or fuel cells a submersible can carry, and human factors constraints: in other words, no one can stay underwater indefinitely; people eventually get fatigued in a submarine environment.

Enter the ROV. With no power limitations (they're electrically connected to the surface and can operate indefinitely), no human factors considerations (if the operators get tired, you just change them out), and no life support restrictions (since no one's underwater, what do you need oxygen for?), they would appear to be the ultimate tool for underwater work and exploration. But this capability comes at a price.

First of all, ROVs always have that damn umbilical trailing out behind them, just waiting to short out, get snagged on aircraft wreckage, or mangled on that piece of rusty steel sticking out from the side of the oil rig that you didn't see when you first dove. On the other hand, as long as that umbilical is attached to the vehicle, it's sort of like walking a dog; it's a lot harder to lose it. Secondly, making all those cameras, hydraulic manipulators, propulsion systems, and telemetry systems work on a reliable basis in the ocean environment is not all that easy. In fact, sometimes vehicles break down for days at a time, or never work at all like they were supposed to. Finally, since no one is actually underwater looking at the bottom, all you've got to rely on with respect to what is actually happening underwater is what you see on the sometimes grainy video image and what other displays you managed to pay someone lots of cash to design, build, and maintain. The other thing to remember is that unmanned submersibles aren't really unmanned; they're just not manned by anyone who happens to be underwater. Also, operating ROVs is not exactly cheap since all those supervisors and technicians don't work for free. They expect (although they rarely get it) big bucks to work 12 hours a day and seven days a week on a ship away from their high maintenance girlfriends or wives. So in the end, the capabilities of an ROV versus a manned submersible is a double-edged sword. You gain something, but it comes at a price, like anything else in life.

One purported advantage of using an ROV is supposed to be safety. Since no one is actually underwater, operations with remote vehicles appear to be, at first glance, somehow safer than sending people into the scary, cold, and dark ocean. But in reality, even remote vehicle operations can be quite dangerous and there have been fatalities. When you consider the hazard of getting equipment weighing up to seven tons in and out of the seas on a rolling and pitching ship in bad weather, it is not all that difficult for a member of the crew to be injured or even killed. In addition, almost all ROVs are powered by high-voltage electrical systems with as much as 3,000 volts running through deck cabling, umbilicals, and the interconnect wiring on the vehicle itself. Finally, just the simple act of even being on a ship is not all that safe, especially when you're recovering interesting little gems like cruise missiles, torpedoes, and mangled aircraft spilling gallons of JP-4 jet fuel as they are dragged on deck.

Who invented the Remotely Operated Vehicle? No one person or organization. Like the airplane, the unmanned submersible was initially developed as a technology partly by the military, in particular the Naval Undersea Research and Development Center in San Diego, California. A few offshore service companies also had a role in remote technology development as well, such as Ametek-Straza (they designed and built the first Deep Drone, SCARAB, and SCORPIO vehicles).

The Navy's first CURV-class vehicles (Cable-Controlled Underwater Recovery Vehicle) were built by a division of the Navy to recover torpedoes lost after being test fired by submarines in the 1960s. By today's standards, CURV was incredibly primitive in that all of the vehicle's functions were independently controlled from the surface using a completely hard-wired umbilical. What this means is that every separate vehicle function was turned on or off by their own individual wires running from the sub to the surface; that's a lot of wires. In addition, they didn't even have a cable storage reel (a rotating drum on which the vehicle's umbilical is stored), which meant that a group of beefy guys had to manually feed the umbilical into and out of the water (which was pretty heavy because of all those wires). This had to have been brutal work for even the strongest man. However, in spite of those limitations, CURV worked reasonably well and was instrumental in the recovery of a hydrogen bomb from deep water off the coast of Spain in 1966 as well as the rescue of two men (Vickers Oceanic's Roger Chapman and Roger Mallinson) trapped 1,500 feet underwater in a Pisces class manned submersible in 1973.

One of the first underwater vehicles to function in the commercial marketplace was the RCV-225 (Remote Controlled Vehicle), built by Hydro

Products in San Diego. The 225 was very small, about the size of a basketball, very maneuverable, and found widespread acceptance in oil fields located anywhere from the Gulf of Mexico to the North Sea. Although the vehicle was painfully underpowered (it used Thimbledrone model airplane propellers to push it around) it really helped prove the capabilities of such vehicles to many a doubting "company man" or drill rig captain who were infinitely skeptical of any new-fangled gadget in *their* oil patch; especially one built by someone not from Louisiana. In fact, the vehicles were used so much in the North Sea that the first thing you were typically asked when trying to get a free-lance job over there was, "How much 225 time do you have?" In short, the RCV-225 was a flying eyeball originally designed to fly out of the torpedo tubes of a nuclear submarine (at least that was the rumor) that changed the whole idea of how to work underwater in the oil field. Divers hated the things because no longer could they get their big depth pay for a quick bounce dive to check out a newly installed piece of oil field equipment. Instead, the company men were asking someone to get a "camera" out to the rig. Both Taylor Diving and Salvage and Martech International were early players in the oil patch with RCV-225s.

But if you could use an ROV to observe stuff underwater, what about doing real work? While the initial emphasis with ROVs was inspection, it wasn't long before a few companies started thinking about mounting some manipulators on a vehicle and pushing divers even farther out of the picture. Both Hydro Products and Ametek-Straza were developing work class ROVs by 1977, the former with their RCV-150 and the latter with the SCARAB and SCORPIO ROVs. But at the time, only one or two RCV 150's were built, two SCARABs (they were financed by a consortium of telecommunications Companies to repair submarine telephone cables), and the SCORPIO was just getting started. It took a small Canadian Company, International Submarine Engineering (ISE) organized by a group of ex-manned submersible pilots to really get things going with respect to doing commercial work with an ROV.

ISE's original vehicle, the TROV (Tethered Remotely Operated Vehicle) was designed due to Jim McFarlane's (the President of ISE) interest in finding the RMS Titanic. By 1978, two such vehicles had been built and sold to McDermott Diving, although they were apparently not used to any real extent (mostly because they didn't have anyone who could keep them running). But when Ocean Systems, Inc. (OSI) placed an order with ISE in 1978 for five ROV systems, a bond was created between the two companies that continues to this day, although OSI has long since been taken over by Oceaneering International, Inc. (OII). Even though OSI requested three work ROVs and two lighter inspection vehicles (called TRECs, or Tethered Remotely Operated Cameras), for almost two years, the TRECs were the systems that seemed to be getting all the work. The powers that be had yet to accept ROVs as a replacement for divers.

The early TROVs were temperamental contraptions but were unique in that they sported soft ballast tanks whereby the surface operator could vary the buoyancy of the vehicle by injecting compressed air into two small aluminum tanks on the front of the sub. In this manner, you could make the ROV ascend or dive by varying the amount of air inside of the tank, reducing the need for a vertical thruster. This was actually a good thing because these vehicles started off using constant-speed variable pitch propellers for propulsion which tended to, uh . . . not work so well. But what was great about the TROVs was that unlike most ROV systems in production at the time, they were designed to be maintained in the field in that the engineers actually considered the possibility that some hapless ROV technician on a rolling deck somewhere might have to dig his hands into the vehicle and fix it when it broke down. This was demonstrated by the fact that the electronics bottles could be swung out from the side of the vehicle to access the circuit cards inside and all of the assorted fluid ports needed to fill the ROV with oil were neatly arranged on a small panel on the side of the frame. Another interesting point about the early ISE vehicles was their habit of using modified off-the-shelf components. The first TROVs were pushed around by sailing boat propellers and used standard Panasonic black and white security cameras mounted inside of custom-made pressure housings. In addition, the TRECs used Evinrude outboard motor props and floor polisher motors stuck in an air-filled housing for thrusters. The only compass TRECs had was a standard divers' wrist compass contained within a camera housing. In other words, to see your heading underwater, you were actually looking at a video image generated by a small video camera looking directly at the compass; a crude method, but it worked. A funny characteristic of the early TRECs was that when the brushes in their thruster motors started wearing out, you could hear interference from the whirring electric motors over your communications headset, thus giving the pilot some unintended audio feedback as to how much thrust was being applied.

In spite of the early 1980's bust in the oil production business, several companies joined the fray in one fashion or another by manufacturing larger, more capable underwater vehicles. British Aerospace had their Consub I and II systems; Perry Submarine, the well-known builder of manned subs, were starting to sell their Recons (a small work vehicle); and several other overseas concerns including Saab Scania (with their Saabsub) were making serious entries into the North Sea oil fields.

The concept of unmanned observation hit the front pages in 1985 when Dr. Robert Ballard discovered the sunken ocean liner Titanic with his Argo towed camera sled during a joint American / French expedition. Although the Argo system was really nothing more than a side scan sonar search vehicle fitted with video and still cameras (and hence not an actual ROV), the fact that someone could observe something on a real-time basis at the near 13,000 foot depths was a big deal in the press as they latched onto the phrase "Tele-presence." This, in spite of the fact that similar operations had been done before (and in considerably deeper water) by the Naval Research Laboratory.

It was also in this period that advances in fiber-optic technology started making it possible to replace some of the copper wires in existing ROV umbilicals with those having fragile glass fibers. This allowed higher-quality video to be transmitted from the bottom and also extended the depth (by virtue of a longer umbilical) at which an ROV could operate. In other words, trying to get usable black and white, real-time video transmitted along a copper umbilical longer than 10,000 feet was not all that practical and color was out of the question. Many of the early fiber-optic ROV umbilicals were notoriously unreliable in that it seemed that all you had to do was *look* at the cable to break the minute glass fibers so critical to vehicle operation. There were also considerable disagreements over how to construct such umbilicals and anyone who created one that worked kept its design a closely guarded secret. The primary problem was mechanical in that it was difficult to mathematically model the physical behavior of a cable as it bent around the large sheaves used to feed it into the water from the surface support ship. In addition, with the ROV driving all over the place underwater, the cable was in constant motion being subjected to all kinds of unknown stresses. If the umbilical bent too sharply for any reason, the glass fibers would break, requiring a time consuming repair job where you had to chop off bad section of cable and remake all of the complicated electrical and optical connectors; a real pain, to say the least.

From the mid-1980s and into the early 1990s, more and more work-class vehicles flooded the market with ISE's Hydra and Hysub vehicles, Subsea

International's Pioneers, more powerful SCARAB vehicles and dozens of SCORPIOs (over 60 were built). The saturation divers who formerly earned over $1,000 a day were seeing the handwriting on the wall as more and more deep diving work was replaced by unmanned technology, which offered the advantages of lower cost, greater personnel safety, and the opportunity for the oil field customers to see for themselves if their newly-installed gadget on the sea floor actually worked.

ROVs that could dive to the deepest depths also sprang into existence with ISE's two Hysub 5000 vehicles and Eastport International's 25,000 foot capable Magellan 725 system. The 725 was an interesting observation in simplicity in that it was built on a shoestring budget in-house by the company using a fiber-optic sonar tow cable rejected by the Navy, as well as bits and pieces from an Ametek Scorpi ROV (Eastport International's Don Dean and Ron Schmidt were key players in the development of the Magellan system). To date, the still operating 725 system has probably spent more time in waters deeper than 10,000 feet than any other system in the world. The Magellan system has since been combined into a dual system of sorts with the Ocean Explorer side-scan offering both a search and recovery capability in one neat package. The Japanese research organization, JAMSTEC, has also created the ultimate deep-diving ROV with their Kaiko vehicle, designed to reach the deepest depths of the Mariana Trench, over 35,000 feet down.

The US Navy continues to operate unmanned vehicles with a new upgraded CURV system capable of reaching 20,000 feet underwater and operated by the Supervisor of Salvage and Diving (SUPSALV). CURV also holds the world record for the deepest ocean salvage after recovering a crashed helicopter from 17,251 feet off Wake Island in the Pacific. But the real workhorse for the Navy is their Deep Drone, which has been used for anything from surveying the Civil War ship USS Monitor to hauling up wreckage from TWA Flight 800. Submarine Development Group I (SUBDEV) also operates several ROVs including a couple of Super SCORPIOs and their deep diving ATV (Advanced Tethered Vehicle).

But even with all of the advances in computers and electronics, ROVs are still temperamental monsters that can behave in the most peculiar manner. Even in 1985, the original SCARAB vehicle had a control system that would not work if it was too cold; sort of its way of saying "I don't feel like working today so go away." During the frigid 10 days the system was transported from St. John's,

Newfoundland to Cork, Ireland to dive on a crashed Air India 747 jetliner, if the surface control van wasn't heated up to at least around 70 degrees, there was no way that vehicle was going to work. The problem is that by the time many ROVs are built and fully operational, electronics technology has moved forward so fast that the vehicle's computerized control system is outdated. Until a few years ago, even the Navy's advanced CURV system used an outdated UNIX-based computer. The system wasn't outdated when the vehicle was designed and built. Unfortunately, by the time you get all of the complicated subsystems designed, built, and actually working, it's already time for a bloody upgrade!

ROVs have been and will continue to be lost at sea, since every time you put something into the ocean, despite all precautions, there is the chance it will never be seen again. The small size of the RCV-225s made them excellent candidates for getting sucked into the bow thrusters on Gulf of Mexico work boats, to be reduced to fragments of syntactic foam (a dense floatation material used on submersible vehicles), electrical wiring, and twisted bits of aluminum. The Navy's Deep Drone once had an unsuccessful encounter with a ship's propeller causing the startled operations crew to ponder their ROV's fate when a section of floatation material bobbed to the surface spelling out "EEP," all that was left of the words "DEEP DRONE." Even the original Deep Drone's replacement, Deep Drone 8000, is toast, having been lost in deep water off of the island of Eleuthera in the Bahamas after getting tangled up in a web of Navy communications cables. The Magellan 725 ROV dove down to the location a couple of years later in a last-ditch effort to get the vehicle back, but the operation was abandoned when it was clear that any operations at the site might result in the loss of a second vehicle. However, there are also success stories, such as the time that the Magellan 725 ROV became wrapped up in rigging on the freighter *Dimetria Beauty* at a depth of almost 10,000 feet. What do you do when your vehicle is stuck in deep water and you don't have another deep-diving ROV to go get it? You build another one. In a matter of weeks, Eastport International threw together another 20,000 foot capable ROV, also built from bits and pieces in their Maryland shop, flew it half-way around the world, and "rescued" the entangled Magellan.

In spite of all the perceived problems related to ROV operations, with an experienced crew, a good vehicle, and attentive ship helmsman, a lot can be accomplished, much of it never having been done before. The real key to any underwater operation, whether it be with divers, a manned submersible, or a remote vehicle, are the people who operate and maintain the gear. They are the ones who make the decisions, sometimes with the benefit of experience, sometimes going by their gut feelings. They are also the ones who have to fix the most complicated equipment under God-awful conditions in the dead of night; even when the right spare parts are at the shop thousands of miles away. It's all about flexibility and innovation; i.e., doing the best you can with what you have to make it all happen.

However, the emergence of ROVs was only part of the solution to any plan to find and recover Liberty Bell 7 as there had to be a way to locate the sunken artifact before hauling it to the surface. Sonar is a system that uses transmitted and reflected underwater sound waves to detect and locate submerged objects and/or measure distances underwater. Before the 1960's, active sonars were primarily used as bottom sounders, where acoustic energy is reflected off the sea floor to measure depth; the submarines used during World War II used passive sonars to listen for surface ships and other submarines (an active sonar transmits sound and a passive sonar only listens for sound). It wasn't until Doctor Harold Edgerton, an MIT electrical engineering professor, started experimenting with sonar that it was developed into a useful search tool. Doc Edgerton, famous for his work with high-speed flash photography, applied the principles of flash tubes to acoustics, creating a side looking sonar that when towed behind and below a ship, created very detailed images of the sea floor. In comparison to today's towed sonar systems, Edgerton's creations were primitive. But they started the ball rolling and in little more than a decade, some elements of the US Navy had the ability to conduct side scan sonar search operations in water depths comparable to the where Liberty Bell 7 was thought to be lost. One pivotal operation was the CHASE X project conducted by the Naval Research Laboratory where they used a towed sonar to locate a Liberty ship, the USS Briggs, that had been purposely scuttled earlier holding concrete vaults of a nerve agent chemical. Working in an area about 120 miles north of the Liberty Bell 7 location, the NRL team successfully located and photographed the sunken ship, also generating considerable oceanographic data that I would eventually put to good use.

However, a sonar capable of operating at 16,000 feet underwater does not solve everything as the operators need to be able to determine what they're looking at on the side scan sonar record. In previous years, people simply looked at a roll of paper that was churned out of the surface thermal printer. What you ended up with were long rolls of thin thermal paper covered with a maze of grayish markings, representing what was on the ocean floor. It took a skilled sonar technician to tell what was geology and what was man-made (it still does for that matter). Also, finding something only 9 feet tall would still have been a tough proposition. However, by the mid-1980's it was possible to locate crashed aircraft in very deep water, such as the Air India Flight 182 747 jetliner (6,200 feet) and the South African Airways 747 lost in over 14,000 feet of water off the coast of Mauritius, in the Indian Ocean. However, that was a whole aircraft. What about finding a singular isolated object in deep water? Something like a Mercury spacecraft?

It took the creation of sonar processing software coupled with desktop computers to make it more feasible to detect objects the size of a Mercury spacecraft. It wasn't that you couldn't do it with the conventional setup, just that it made picking out small targets against the background far easier. The raw sonar data routed though these data processors is the same data turned into a hard copy by the thermal printers. The difference is that the processing software gives the operator the ability to manipulate the data by using color (among other things) to represent the hardness of sonar targets. What this means is that hard metallic objects (i.e., things that reflect a lot of sound) can be displayed in red, while the more benign background (i.e., bottom sediments that soak up much of the sound) can have a soft blue color (depending upon what color palette the operator has selected). The best part is that this processed sonar data can be coupled with a satellite navigation system allowing the sonar operator to quickly determine the geographical location of any indicated sonar targets.

How to actually lift Liberty Bell 7 was another problem that needed to be solved. Depending upon the size of the object, there are only two ways to recover something from the bottom of the ocean: You either have to increase the target's buoyancy (i.e., some sort of lift bags) or simply attach a line to the object and raise it to the surface.

There had been several important salvage operations when Liberty Bell 7 sank, but none of them at comparable water depths. The US Navy submarine USS Squalus was raised from 243 feet of water in 1939 using a series of massive air filled pontoons positioned earlier by divers and cranes. While the operation was a success, similar techniques would have been impractical for raising Liberty Bell 7. This is due to the tremendous ambient pressure at depth, in this case, about 16,000 feet. At that depth the pressure on every square inch of any object is 7,127 lbs. To try and inject air into any lift bag or pontoon under those conditions would have been all but impossible. Lifting the spacecraft using some sort of line might have worked in 1961, but just barely. A standard 3/8" diameter galvanized steel wire rope has a breaking strain of almost 11,000 lbs.; certainly enough to raise Liberty Bell 7 which weighed a little less than 2,000 lbs. in the water. But a length of wire rope long enough to reach the capsule weighs over 4,000 lbs. by itself. A kevlar line would have been great as they're about as strong as steel and don't weigh anywhere near as much as steel wire ropes. However, in 1961, there was no such thing as a kevlar liftline because Dupont had yet to invent the substance.

Overall, when Grissom's Mercury spacecraft disappeared beneath the surface of the Atlantic Ocean on July 21, 1961, our ability to operate in such water depths was extremely limited at best. This was before the search for the submarine USS Thresher, before a hydrogen bomb was recovered off the coast of Spain, and before the CIA raised part of the Russian submarine K-129 in the Pacific Ocean. For all intents and purposes, Liberty Bell 7 might as well as have been on the Moon for all anybody could do about it.

Chapter 6 – Investigation
Unlocking NASA's Archives

Knowledge is of two kinds. We know a subject ourselves, or we know where we can find information upon it.

— Samuel Johnson

By the year 1985, a lot had happened in the world since Grissom's 1961 sub-orbital space mission. Both John F. Kennedy and his brother Bobby had been assassinated, six presidents had gone through the White House, and the submarines Thresher and Scorpion both sank with a loss of all hands. Virgil I. Grissom, referred to by some as "the hard luck astronaut," died on the launch pad in the Apollo I Command Module along with Edward White and Roger Chaffee. Fellow astronaut Neil Armstrong took the first steps on the Moon, NASA's Space Shuttles started flying out of what was now called the Kennedy Space Center, and small hand held calculators were invented. People started wearing digital watches, computers had gotten so small they fit on a table top, and I was trying to exist as an out of work ROV pilot living in Colorado Springs, Colorado.

By then I had spent about 11 years working in the subsea field in one fashion or another. I started out building ship fenders at $3.50 an hour for Ocean Systems, Inc. in a fabrication shop near Dulles Airport on the outskirts of Washington, DC. When I got started in the "business" in 1974, my mother was still living and we shared a house in Vienna, Virginia. After about a year and a half, I graduated from ship fenders and started building deep diving systems, mostly diving bells and deck decompression chambers, the steel prisons used by divers after spending several hours deep underwater. When I got going in the field, there was no such thing as a commercially operated Remotely Operated Vehicle. The Navy had one but that was about it.

Some of my earliest thoughts about the ocean in my hometown of Oakland, California, were about undersea exploration. While living at the Presidio Army Base, near San Francisco, I clearly remember riding in the back seat of a car and seeing the mast of a sunken ship sticking out of the water in a rocky cove near the main road into the installation. Even though I could not have been older than 5 or 6, I imagined what it would be like to explore the ship underwater. While I visualized a pirate ship, the reality was probably a local fishing boat. Eventually, the mast rusted away and the object of my childhood fantasies ceased to exist. However, my interest in the underwater world stayed with me.

As a youth growing up near St. Louis, Missouri, I suppose you can say that I had two passions: space flight and undersea exploration. My grade school teacher used to rip drawings of rockets off my small desk and crumple them up in front of the whole class. I mixed my own black powder, built small rockets, and was lucky I didn't burn our house down. Since my father was a career Army officer, we never spent too much time in one place. However, it was in Florissant, Missouri that I rode my bicycle with a Hitachi transistor radio stuck to my ear listening to the countdown of Alan Shepard's Mercury mission. In general, all of the Mercury astronauts were big heroes at the time due to the constant exposure they received in LIFE magazine. It seemed like every week there was a new article about one of the Mercury guys. While Shepard and Glenn were certainly the most famous to me, I remember being taken by Grissom for no special reason. Maybe it was the way he looked or that he didn't appear to seek out the limelight. However, he was a central figure to me.

When not trying to blow myself up, I also "designed" submarines. I would do the most detailed drawings of small submersible vehicles that I fantasized I could pilot deep underwater while searching for some lost ship or sea creature. My creations were supposed to be made from surplus aircraft wing tanks, car batteries, electric fan blades (for propellers), and even had provisions for complicated ballast tanks, valves, hatches, and view ports. I think I was 12. I remember reading an article about John Perry, the founder of Perry Submarine, in Popular Science magazine showing off his new miniature submarine and even wrote him a letter asking for technical information on his designs. Surprisingly, he actually wrote back to me enclosing a brochure showing the design of a commercial submarine having an area where divers could be locked out. While other kids were playing baseball, I was devouring books about John Holland and Simon Lake, learning as much as I could about how they built their submarines.

When I was a teenager in high school, living near Washington, DC, my father was working for the Assistant Secretary of the Army at the Pentagon. I was a typical 1960's kid, trying to have long hair without getting suspended from school and playing in a variety of rock bands. But my father did a good thing for me: one night he came home from work with a small booklet he must have picked up at the Pentagon. It was called "Undersea Vehicles for Oceanography," and described all of the latest technology for exploring the deep ocean. If I had been smart, what I would have done was get good grades in high school, an engineering degree, spend a few years in the Navy in submarines, and then work at my fantasy job – being a submersible pilot. However, like many teenagers, I was stupid. I barely got through high school, dropped out of college, and started everything from the bottom: the hard way.

I also have to say that I had good parents in that they always supported me with what ever I wanted to do. My father was busy much of the time, flying light aircraft out of Chrissy Field at Presidio, helicopters in Korea for a MASH unit, helping map Alaska with Dehavilland L-20 Beavers, and doing two tours in Vietnam, shooting up the landscape with helicopter-mounted rockets. But even though I was the son who was always getting in trouble, I knew that he wanted the best for me. In many ways, I felt like the "black sheep" of the family in that while my brother graduated from college, went into the Army, etc., I did everything wrong. I had long hair, listened to rock music, had to go to summer school, and dropped out of college. I made a lot of mistakes. But I learned. My mother was equally supportive of me and really did a good job of raising my brother and I. Just like Betty Grissom, she had to run the household when dad was away and do it with two kids running around and I know it wasn't easy. When I was 19 and just out of school, my parents let me go to England and learn how to drive Formula Ford racing cars at the Jim Russell Racing Driver's school. I still cannot believe they let me do that at such a young age. But they did.

After holding menial jobs working on the loading dock for a department store, as a landscaper, a failed security guard, and a rock driller, I finally made my entrance into the underwater business in August of 1974 with Ocean Systems. I had no formal technical training other than a machine shop course at a local community college and what I knew from building and racing Formula Vee race cars. I was 24 years old.

In the years between 1974 and 1985, a lot happened to me and I received a fair amount of experience with underwater vehicles. I did an aborted job in Taiwan doing coral harvesting with the SCORPIO I ROV (our Chinese client ended up stealing our ROV), inspected pipelines in the Gulf of Mexico with TROV S-4, did a bottom survey of part of the Mediterranean, inspected and cleaned Phillips Petroleum oil production platforms in the North Sea, and even buried telephone cable with SCARAB II for AT&T. I also went back to college and got a two-year degree in Electronics Engineering Technology, my first formal technical training. Overall, I had a good mix of experience with doing underwater work. I felt I was as good an ROV pilot as most guys, not a bad technician, and a competent supervisor at times. However, by 1985 I was getting bored with doing the same kind of underwater work all the time. I used to read books and see films about people like Cousteau doing things underwater that appeared to be more interesting. I wondered, "How do you get to do that kind of stuff?" It all sounded like a lot more fun than spending hours looking at a rusty gas pipeline just offshore Venice, Louisiana.

As a result, I started thinking about the things that could be done with underwater technology that might be more personally rewarding. The idea of trying to be a public figure really wasn't part of it. What I wanted was to do something more creative with the technology; something I could organize and take credit for. During the typical offshore job you're always working for someone else, whether it be an oil company, telephone company, or engineering firm. I was interested in doing something that *I* felt was worthwhile with the underwater vehicles I had worked with for so many years. I wanted to have some fun with ROVs.

At the time, I was living with my father in Colorado Springs, looking for some freelance ROV work, and basically trying to put my life back together. I had little money, but a lot of ideas. I started thinking about things that had been lost in the ocean. *Targets.* Sunken objects that would be interesting to find and explore. I came up with two possibilities: the Titanic and Gus Grissom's Liberty Bell 7 Mercury spacecraft.

I knew that Ballard was working on the Titanic and really had no idea how to organize such a project. I wrote a proposal of sorts, but didn't know where to send it. Who would be interested in searching for an old sunken ocean liner? I had no idea. But I wrote an article that was published in a diving magazine called Underwater USA, which outlined a plan for how the Titanic might be found and what kind of condition it might be in. Surprisingly, when the Titanic was eventually found, my predicted location was off by only three miles.

I also started investigating the loss of Liberty Bell 7. When I began, I didn't know how deep the water was where it sank or even if it was in the Pacific or Atlantic Oceans. But I remembered that it had been lost and not ever recovered, as far as I knew. I also figured out that there's a reason things stay lost for decades: *they're either not worth going after or they can't be found.* I only had one book that mentioned Grissom's flight in any detail, "Appointment on the Moon," by Richard S. Lewis. It didn't have much information about Liberty Bell 7's loss, only saying that, "Liberty Bell 7. . . sank to the bottom of the Florida Trench, 2000 fathoms down, to join the fabled hulks of Spanish treasure ships and become the first sunken spaceship." Two thousand fathoms equals 12,000 feet underwater. Where the hell is the Florida Trench? Must be in the Atlantic for sure. I also had a paperback copy of "We Seven," which I knew had some information on Grissom's flight. However, it didn't say anything about the water depth or where the capsule was lost. I also vaguely remember finding another book saying that Liberty Bell 7 sank in 16,000 feet of water. Man, that's deep! I couldn't do any research on the Internet because in 1985, it didn't exist.

Well, I figured, there had to be ships and helicopters around the capsule when it sank, right? So, there should be some record of exactly where it sank. I later learned that "exactly" is a relative term.

Two things happened in 1985 that deeply affected me. First of all, after a two-year struggle with Lou Gehrig's Disease (Amyotrophic Lateral Sclerosis), my mother died. It had been a difficult experience for me to take care of her in Virginia before she moved back to her mother's house in Oakland, California. I moved there with her but after a few months, it was obvious I could not stay permanently due to tensions within mom's family. The sorrow I felt is something that only someone who has lost a parent can understand. She raised me. She helped me. And now she was gone. Fortunately, my brother Chris was there to handle the funeral arrangements because I was a basket case. I still miss her.

Shortly after mom passed away, I got a call from a former coworker at Eastport International asking me to help out in some short-term ROV work with SCARAB II off the east coast near Providence, Rhode Island. I went. It was good for me in that it kept me from thinking of my mother. During the job, which was only supposed to last for about 10 days, an Air India 747 jetliner crashed off the coast of Ireland killing 329 people (the loss of Air India Flight 182 remains the worst air disaster at sea). All of them were Canadian citizens and since we were working on a Canadian Coast Guard ship, the CCGS John Cabot, it was not long before we were steaming across the Atlantic to help salvage the crashed jetliner and figure out why it went down.

The Air India operation remains one of the best jobs I've ever worked on. The SCARAB ROV worked like a champ in the 6,200 foot deep waters breaking several endurance records for underwater vehicles along the way. On one dive, we worked in waters a mile deep for 138 hours – almost six days. No one had ever done that before. It was also a rewarding experience working on such a high-profile operation, as the salvage was big news in Ireland and the rest of Europe. None of the numerous underwater jobs I'd worked on before had ever received any publicity, except within the industry. When you tell people about these projects, it's as though they don't have a clue what you're talking about. But when you describe something they've read about or seen on television, somehow it all clicks in their brain. This was something people would hear about. In the end, I spent six months working on the John Cabot, seven days a week, and 12 hours a day. I was on that ship from June 9th until November 18th, 1985, until I finally got sick from exhaustion and asked to be sent home. I did every trip to the Air India work location except for the last one. I was burnt out.

However, in addition to giving me some valuable salvage experience, the Air India job did something else. It told me that if Liberty Bell 7 could be found, it could be recovered using the same techniques we used to raise wreckage from the 747 jetliner. There was no doubt in my mind. In about 13 hours, we could make three dives to the bottom in 6,200 feet of water and recover aircraft wreckage while doing a four-point lift in the process. If an underwater vehicle capable of reaching Liberty Bell 7 could get near the Mercury capsule, there was no reason we could not do the same thing. Assuming, of course, that we had over 16,000 feet of kevlar recovery line to play with and an ROV that could dive that deep.

In 1985, there was no such thing as an ROV that could reach 16,000 feet underwater. In fact, there were only two submersible vehicles in the world that could get there, both of them manned systems: the US Navy's Trieste II and Sea Cliff DSV. That was it. IFREMER's Nautile had yet to be built, JAMSTEC had not even designed their Shinkai 6500, and the Russians had no Mir submarines. There were no remote vehicles for 6,000 meters because no one had figured out how to design an optical fiber umbilical that worked. The existing cables using coaxial wiring for video could not send real time video over that length of copper wire. However, while I had no way of predicting the future of underwater technology, I knew that *if*, and that was a big if, you could get some sort of underwater vehicle near Liberty Bell 7 it could be raised.

Before going back to Colorado, I stopped off in Washington to get paid by Eastport and also visit with Bill Wood, an old high school friend living in Arlington, Virginia. For a few days, I used his condo as a base of operations and hit the NASA History Office, seeking information on Grissom's flight. They had some newspaper clippings and a couple of declassified NASA documents, but not much. I photocopied some pages and figured I had done my research. Wrong. I also set up a meeting with some people from the National Air and Space Museum, after telling them about my interest in recovering Liberty Bell 7. While the meeting with Brian Duff and Lin Ezell went well enough, after sending them a proposal about the project, they expressed little interest in such a recovery effort:

> ... it's our general opinion that a project to dive for and recover the Gus Grissom Mercury Space Capsule is not something for the National Air and Space Museum to sponsor or even to become involved with in a principle way... while we are interested in the preservation of key development artifacts, the heart of your project is the finding and recovering of the spacecraft and diving and salvage is not our main mission.
> — Brian Duff, National Air and Space Museum, December 14, 1985

From 1985 to 1986, I drifted with no specific direction. I didn't know what to do with myself. I left Colorado and moved to Los Angeles with all of my belongings that would fit in my Saab. I went there for no particular reason. Maybe I thought I could take up music again. Maybe I could find a job working with ROVs. I simply didn't know what I wanted and sat alone in a rented Oakwood apartment burning up money and trying to figure out how to go after Liberty Bell 7.

However, I soon managed to get on another salvage job when the Space Shuttle Challenger exploded. In some ways, doing underwater salvage operations after aircraft accidents is like being an underwater ambulance chaser. We only went out after something to pick up the pieces following a disaster. Even so, it was important for NASA to figure out why the Shuttle blew up, and I spent two months working out of Port Canaveral, Florida on the salvage ship Stena Workhorse, a massive Swedish heavy work boat. The operation was a real grind, mostly due to the numerous technical problems we had with the Gemini ROV. In six weeks, we repaired its electrical umbilical a staggering 32 times and even replaced the whole thing four times; not a good record. But while I was in Florida and on one of my rare days off, I visited the archives at the Kennedy Space Center and collected a little more data on Liberty Bell 7.

While I got no closer to finding Liberty Bell 7, I finally found a full time job in late 1986 working for Eastport International with the SCARAB II ROV, the same one we used in Ireland on the Air India job. I also did some hard thinking about Grissom's capsule and decided that there were three questions that had to be answered before I could even say the project was feasible:

1. Was there sufficient navigation data available to develop a search area having a high probability of containing the spacecraft?
2. Would the spacecraft be in good enough condition to warrant recovery, taking into account the materials used in construction and the deep ocean environment?
3. Was underwater technology sufficiently advanced to make the project feasible, from the technical standpoint?

Until I could answer those questions, there was no point in asking anyone for money because I didn't even know if the project could be done. I needed to do a research program. At that time, there was no doubt in my mind that once I could prove the capsule was intact and capable of being raised, that finding the money to do it would be a breeze. I was very naïve. It was in 1986 when I met Gregg Linebaugh, a space enthusiast and artifact collector. Gregg had seen an article in Space World, a magazine published by the National Space Society, and was interested in helping out. Even though I doubted his claims to begin with, he said he knew many of the astronauts from the Mercury, Gemini, and Apollo programs and was willing to see what they could do to help.

It was about this time that I first met Max Ary, the President of the Kansas Cosmosphere and Space Center in Hutchinson. Max had managed to turn a small Midwestern planetarium into a well-respected space museum and also developed a reputation of being a space "pack-rat" of sorts; he had the uncanny ability to find discarded space hardware in the most unusual places. He also admitted to considering the possibility of recovering Liberty Bell 7 and recounted a story of how he had once called the US Navy to get information on whether or not the spacecraft could be raised. After the Navy stopped laughing, they told Max to forget it: the water was way too deep in that area.

I also started collecting more NASA and US Navy documents, in particular, the Post Flight Memorandum published by the Space Task Group in 1961 after the capsule was lost. I also found copies of the deck logs for the Navy ships assigned to the capsule's recovery, in the hopes that there might be some good navigational data in their deck logs. But amazingly, the USS Randolph, the prime recovery ship, didn't bother to take an 0800 hours fix, which would have been just after Grissom landed. As a result, the Navy deck logs were not much help. In addition, I finally received some clarification on who actually owned Liberty Bell 7, taking into account it had been sitting on the bottom of the ocean for over two decades:

> Liberty Bell Mercury capsule was built with funds appropriated by the Congress to the National Aeronautics and Space Administration... Neither the United States nor NASA... has ever stated that the capsule is excess property and has never relinquished ownership of the capsule. Therefore, if the capsule were ever recovered, NASA would claim ownership of the capsule because it has never ceased to be U.S. Government property.
> — Helen S. Kupperman, NASA, Special Assistant to the General Counsel, December 3, 1986

Finding information on the Mercury Redstone No. 4 mission was hard, because none of the documents were in one place. A handful were at the NASA history office, others at the Kennedy Space Center Archives, even more at Rice University, and a scattering at the Marshall Spaceflight Center in Huntsville. This took many months, even years, to do but once I had the documents, I had an even greater problem: *I had to actually understand what they meant.* This was not easy as my expertise was in underwater operations, not ballistic trajectories. I had to try and learn what the difference was between landing points created by things called FPS-16, I.P. 7090 Integrated, Mils (Sofar), MCC at loss of signal, and AZUSA MK II. I had no idea what this terminology meant. How accurate were the locations created by these assets? Were they radars? Or something else? The declassified NASA documents were useless to me unless I could figure out what the information meant with respect to mounting an underwater search operation.

However, one thing burned its way into my mind. In NASA's Postflight Memorandum, they listed two locations in a numerical table, one called "Planned," and the other called, "Actual." If nothing else, I knew what that meant. The "actual" location listed by the Space Task Group had to be their best estimate of where Liberty Bell 7 actually landed. However, even with that, it was a matter of opinion. What I figured out was that many people worked on the trajectory of Liberty Bell 7, not just the Space Task Group at Langley. Engineers and scientists at both the Marshall Space Flight Center and the Goddard Space

Flight Center created their own technical documents detailing where they thought the spacecraft was when it landed. Which location was I supposed to use? Early on, I decided to baseline my search area as a box eight miles square on each side; Liberty Bell 7 had to be in there somewhere.

As I continued digging through photocopies of declassified NASA documents, Gregg introduced me to Lt. General Thomas P. Stafford, who was one of the few astronauts who actually assisted me in these early efforts. Where I had tried without success to get manufacturing drawings of Liberty Bell 7 from McDonnell Douglas Space Systems, all Stafford had to do was call up his buddy John F. Yardley, the former Chief Engineer for Project Mercury; in about two weeks, I had all the drawings I needed in addition to a brief engineering study on how much the capsule might weigh when lifted out of the water. The thing about Stafford was that he fooled you. Tom had a lot more political savvy than he let on, was a hell of a smart engineer, and also had a serious hand in the development of the Space Shuttle. Not many people knew that. Plus, I felt his interest in Liberty Bell 7 was genuine.

I continued working at Eastport International throughout 1988, only getting to do one recovery job, picking up an F-16 fighter aircraft off the coast of Japan. It wasn't long before I was enlisted to help build a replacement for the US Navy's Deep Drone ROV, to be called Deep Drone 8000. My problem was that with me being stuck at the shop building the new vehicle, I wasn't getting any field time (i.e., overtime), which is the only way I made any money. Consequently, I was not making enough to live on and started looking at other options.

> *Thank you for your letter... concerning a project which you feel I may have interest... Unfortunately, I do not have any funds available for support of the Liberty Bell search and recovery.*
> — R.E. Turner, Turner Broadcasting System, December 17, 1986

I stuck it out with Eastport until 1987, when I quit and started working for Oceaneering Space Systems, a new division of Oceaneering established by then future NASA astronaut Mike Gernhardt. I also had an ulterior motive in taking the job. I figured that if I didn't go offshore anymore that I could devote more time to the idea of finding Liberty Bell 7. While working on the Space Station Freedom program was enlightening, it educated me for the wrong reasons. From what I saw, the government was wasting millions of dollars on a design that would probably never be built. In addition, engineering on the Space Station program was more linked to politics and turf wars between NASA centers than technical principles and that's no way to run a research and development program. Unfortunately, that's what they tried to do and the five long years I spent on the space program were for the most part, a waste of time.

One thing I did, however, was get a better grasp on how Liberty Bell 7's landing point was determined. In 1961, the pre- and post-flight splashdown positions were created using several assets. First of all, before the Redstone was ever launched, NASA had a rough idea of where Grissom was supposed to land based on the specific impulse of the Redstone, burn time, and the effect of the capsule's posigrade rockets on the actual trajectory. To actually track the capsule in flight, I learned, they had an overlapping series of precision tracking radars (i.e., FPS-16) which fed real-time tracking data to the two massive IBM 7090 mainframes (newly transistorized replacements for the old IBM 709 computers) they had running in parallel at Goddard. I believe that after they crunched the numbers at Goddard, the data was fed back to the Cape to run the real time plot boards. They also had *integrated trajectories*, which were based on real time sensor data fed from the Redstone back to the Cape via a telemetry link. This way, they could take measured inertial data from the booster and calculate where Grissom was going to land before he actually did. They also had another radar called the AZUSA MK II; I never really figured out what that one did. The FPS-16 radars, which are still in use today, were very accurate. In fact, the published ranging accuracy of that radar was about plus or minus 25 meters.

My problem was that the FPS-16 radars did not track the spacecraft all the way down to the ocean, but only until Liberty Bell 7 dropped below the "radar horizon" at an altitude of about 11,000 feet. From reading the Post Flight Memorandum and several trajectory reports, it was obvious to me that the last NASA radar to track the capsule was the one based at Grand Bahamas Island (GBI, the 3.16 Radar); it was this radar that was used to generate the fix identified as "GBI Radar" in the documentation. I also tracked down the German engineer who did the acceptance tests on the NASA GBI radar, Mr. Peter Hoffman-Hayden, who gave me valuable information on the real-world accuracy of that particular radar. I also learned that the capsule was actually tracked using two radar transponders (C and S band) which made the location even more accurate. A big question for me was whether or not NASA took into account the capsule's drift on main parachute with respect to their radar positions. After considerable digging, I concluded that the radar fixes were where they had the capsule just before they lost it on radar.

I determined the capsule's drift on main parachute by using the winds aloft data for the day of the flight as measured at Grand Bahamas Island; I used this information to plot the actual track of the spacecraft as it descended to the ocean on its main parachute (it drifted to the northwest about 1.3 miles). Taking into account the accuracy of the radar fixes and possible drift error on main parachute, in theory, in 1961, NASA should have been able to define the landing point of Liberty Bell 7 within a nautical mile.

> *I have contacted individuals at the Gannett Foundation and they do not feel that they are in a position to make any kind of commitment to a project of this type... Gannett's major grants are dealing more on the "human need."*
> — Frank Vega, Florida Today / USA Today, October 16, 1987

While working at Oceaneering Space Systems, I learned that they were planning to do some deep water sea trials using the Gemini ROV we used on the Challenger salvage; it had since been updated and now had a 15,000 foot depth capability. I made the suggestion: Why not add a side-scan sonar to the trial and use the opportunity to look for Liberty Bell 7? After considerable back and forth with several Oceaneering vice presidents, they decided to give it a try using Steadfast Oceaneering's Deep Ocean Search System (DOSS).

One issue that had to be addressed before Oceaneering would even attempt to look for Liberty Bell 7 was ownership. In order to justify the cost of the operation, we were hoping to get title to the capsule once it was recovered and donate it to a museum for a corporate tax write-off. Even though NASA maintained that the spacecraft was still government property, the Smithsonian Institution had the right of refusal on all space hardware "excessed" by NASA. Given that the National Air and Space Museum had expressed little interest in Grissom's craft, I managed to get them to officially decline Liberty Bell 7 as an acquisition. Both NASA's Office of the General Counsel and the General Services Administration (GSA) approved the paperwork needed to allow NASA to transfer title to Oceaneering, assuming we were successful. The process took six months and a phone call from Tom Stafford to NASA to make it all work.

By early 1992 I had assembled a detailed operational plan and along with some Oceaneering navigators, developed a geographical plot of all of the historical tracking data as well as where we planned to search for the spacecraft. If I had a beef about how the project was organized, it was that they put all of the operational control into one of their supervisors. I felt that I should have more say about what was done given my knowledge of the flight. After arriving at the splashdown location, we only did two track lines, both of them right on top of each other. After sighting two small targets, our "supervisor" declared that the capsule had been found and the search was over. I was flabbergasted. How could you do two track lines and call it a search? You didn't. I had to admit, that the targets looked good in some respects, not so good in others. The size was about right for the bigger of the two, but it didn't seem to have enough shadow for something as tall as a Mercury capsule. Plus, there were two targets only 50 feet apart on the bottom (the smaller one was about the size of the hatch). I couldn't see any way the capsule and the hatch could land so close together given what I knew. My calculations suggested that the capsule's hatch took as much as 30 minutes longer to hit bottom than Liberty Bell 7 (the spacecraft took about 1 to 1½ hours); as a result, the hatch would have been farther down current from the capsule.

After waiting for over a year, we did a series of test dives at the spot (using the Magellan 725 ROV) and discovered that the two targets were aircraft wreckage; apparently, sometime after the capsule sank, a twin engine private aircraft had crashed at my best estimate of where the spacecraft sank. What were the odds? Astronomical. However, for me, I saw it as an even bigger problem. Now, at an isolated spot of the ocean over 100 miles from the nearest land, I had a search area littered with unwanted targets that would make looking for the capsule that much harder. It was as though someone took a golf ball, tossed it into a bucket with about 100 other balls, and then asked you to find it when they all looked almost identical.

> *Mr. Perot certainly appreciated the information concerning the recovery of Liberty Bell 7. He sincerely regrets, however, he will be unable to participate in this worthy endeavor.*
> — Sally Bell, Secretary to Mr. Perot, April 11, 1988

My situation at Oceaneering Space Systems finally came to a head when they decided to shut down our office due to a restructuring of the Space Station program. Except for me, everyone at the office was offered jobs at Oceaneering's Advanced Technologies division in Upper Marlboro, Maryland. I was given three days notice, two weeks pay, and promptly shown the door. So far, Liberty Bell 7 had been nothing but bad luck for me. Now I was out of work, just before the holidays in November of 1993 when the offshore season was winding down, and forced to try and find a new job. I collected unemployment and tried to find something, anything, that would pay me enough to keep my one small piece of real estate, a condominium in Arlington, Virginia.

The idea of finding Liberty Bell 7 became a hobby for me. Other people collect stamps. I analyzed flight data and construction drawings for a 1961 spacecraft lost in three miles of water. Even with the personal problems I was experiencing, I managed to learn more about the Mercury spacecraft. I delved into the materials used to build the capsule and what would happen to the spacecraft when it was subjected to the tremendous pressures on the bottom of the deep ocean – over 7,000 lbs. per square inch, in this case. What I learned was that the basic structure of the capsule could not have been made of better materials. Almost all of the load bearing structures were made of either pure titanium, or titanium alloys, making them virtually impervious to salt water corrosion. The outer corrugated shingles were made from a nickel-steel alloy, called René 41, which was almost as good as the titanium in the ocean environment.

In addition, from the materials standpoint, the best part was the heat shield. What I found out was that the first two suborbital missions, Shepard's and Grissom's, used spacecraft fitted with beryllium *heat sinks*, as opposed to the fiberglass phenolic *heat shields* fitted to the capsules that actually went into orbit. When different metals are coupled and submerged in salt water, the galvanic characteristics of each individual material cause them to act like batteries, where a potential is created between one material and the other. In the case of Liberty Bell 7, the beryllium heat sink would be extremely anodic (it's near magnesium and zinc on the galvanic scale of metals), as opposed to the cathodic characteristics of the titanium. This meant that the heat shield would act like a sacrificial anode when Liberty Bell 7 was submerged in the ocean. In other words, instead of a spacecraft, Liberty Bell 7 was a 1200 lb. titanium structure connected to a 340 lb. beryllium anode by 48 stainless steel straps and cables. That, in combination with the low oxygen and cold temperatures at depth (about 36 degrees F), almost guaranteed that the capsule should be intact and strong enough to be lifted. I also figured that all of the lettering on the exterior would be readable as well, mostly due to the cold temperatures, materials they were painted on, and lack of ultraviolet light (or any light, for that matter).

It is with regret that I must inform you that your request of 12 April for seed money to recover the Liberty Bell 7 has been turned down by the McDonnell Douglas Foundation...Your proposal for this effort is certainly ambitious and exciting and we wish you well in your endeavor.
  — John F. Yardley, President, McDonnell Douglas Astronautics Company June 24, 1988

What about the pressure? Wouldn't that have crushed the capsule after it sank? No. I did an extensive study of the various subsystems contained within Liberty Bell 7, looking at them solely from the standpoint of a mechanism that was dropped in the ocean. The basic structure should be fine, except that all of the air-filled dimples used to add strength to the titanium sheet would be squashed flat. The two helium low pressure spheres would probably crush, but I was reasonably certain that the high pressure oxygen bottles would survive, due to their high internal pressure rating. The implosion of any air-filled cavities on the capsule might create some pretty powerful shock waves, especially underwater. I also thought that all of the gauges were not hermetically sealed and would simply flood when the capsule sank (they didn't). Some of the stainless steel and aluminum tubing might be crushed as well, but overall, Liberty Bell 7 should be in good enough condition to raise, and be worth restoring. On a flight from Washington to Colorado and with nothing else to do, I even calculated the submerged weight of the capsule based on the materials used and their respective weights: slightly more than 1,500 lbs. and well within the capabilities of a kevlar recovery line. Overall, by 1994, I had easily answered two of the questions I needed to answer: the capsule could be found and it was worth recovering, if underwater technology was up to the task.

The approximate weight of the capsule is 2,100 pounds plus the water. The hatch is removed from the capsule and the estimated amount of water in the cabin, when hoisted with the recovery section up, is 2,000 pounds (30 cubic feet). If the impact skirt is still attached the amount of water the skirt will contain when hoisted to the surface is 8,000 pounds (127 cubic feet). The skirt contains holes and the water will drain when lifted slowly.
  — R.A. Graham, Section Chief – Design, McDonnell Douglas Astronautics February 5, 1987

I finally found some offshore work with Margus Company, a small firm established by a couple of old friends I met while burying telephone cable with

SCARAB II. I spent some time near Halifax, Canada, using AT&T's Sea Plow V to install about 179 kilometers of optical fiber cable in the Atlantic near Halifax. Once again, we used the Cable Ship John Cabot, which by this time, had been sold to Teleglobe of Canada and painted a horrible puke green color. What got me was that while we were laying the cable from our ship we could actually make telephone calls home via this link to Halifax.

After Canada, I started working in the south of France and off the coast of Sicily using an ROV Margus had purchased from a US Marshal's sale. It had been sitting for a long time and took quite a bit of work to get going. For almost two years, I shuttled back and forth between Bethesda, Maryland (where I was living at the time) and either La Seyne, France or Catania, Sicily. The work was extremely difficult due to the numerous technical problems we had with the ROV, which had some unusual (and unreliable) design features. In all my years working with ROVs, I had never before seen propulsion units actually explode and vehicle components crushed by water pressure. On that job we saw all of that and more.

Given that I was doing the work in the Mediterranean on a free-lance basis, there was little security. I never really knew when I might be working. As a result, in 1996 I once again accepted a position at Eastport International, which had by then been purchased by Oceaneering International and renamed their Advanced Technologies Division. It was also during this period that I actually finalized a plan for finding and raising Liberty Bell 7.

I had known about Oceaneering's Magellan 725 and Explorer 6000 system for many years, especially after we used that vehicle to dive on the two sonar targets that were not Liberty Bell 7. What I liked about the equipment was it was simple in design and operation. In other words, while the Magellan ROV didn't have all the bells and whistles that other systems have, the thing worked; and pretty damn well as far as I could tell.

The Ocean Explorer 6000 side scan sonar, I thought, would be ideal for finding Liberty Bell 7. This was mostly due to its unique ability to search an area with a wide swath and still be able to detect small objects on the bottom in waters down to 6,000 meters. From what I could tell, with the Ocean Explorer, we could search the bottom while carving out a path 1,000 meters wide and still be able to detect objects on the bottom as small as ½ meter in size. Not only that, but we could search simultaneously on two frequencies (33 kHz and 100 kHz) and flood the bottom with acoustic energy. This was important, because it meant that we could examine my search area in about a week, as opposed to other higher frequency sonars, which could, easily take twice as long. In addition, the ROV and search vehicle operated as a dual system in that both the Ocean Explorer and Magellan were taken to sea. Once you had finished searching an area, you could switch over from the sonar to the ROV and confirm what was down there. This was an important capability.

> Mercury Seven Foundation recently had a Board of Directors meeting in Orlando, Florida... During this meeting the Liberty Bell 7 recovery project was discussed in detail... it was their unanimous opinion that unless it could be supported or sponsored by an organization like National Geographic... they should not be involved. Therefore, the Mercury Seven Foundation will not commit any effort to raise funds for the project.
> — Lt. General T. H. Miller (USMC Retired), Mercury Seven Foundation
> June 12, 1989

At Oceaneering, I started off helping build the Sea Tractor, a massive tracked underwater vehicle designed to bury communications cables for the US Navy. But in my mind, I continued formulating how the Ocean Explorer and Magellan could be used to find and raise Liberty Bell 7, assuming I could find someone to pay for it all. In the over ten years I had spent trying to get the project off the ground, I had spent thousands of dollars of my own money, been dumped by numerous girlfriends (who got tired of hearing about Mercury capsules and Gus Grissom), and felt like I had become somewhat of a joke to people. In addition, not one person had given me a dime for the recovery. *No one.* Everyone thought it was a good idea but no one wanted to pay for it. No one had the balls to take the risk. My peers were dubious about finding the capsule in such terrain. Even so, I could not let go of the idea. I *knew* that if we could find the thing we could raise it. I *knew* that. The question was, did I know what I was talking about or was I crazy? Was it *really* possible to find such a tiny object whose location was determined using 1961 technology? Was the spacecraft *really* worth recovering? Would it hang together during a three mile trip to the surface? Even I didn't have the answers to those questions. No one did.

I also started thinking about the step by step procedure it would take to actually lift the capsule to the surface. I figured that it would take about three dives to do the job. On the first dive, the Magellan ROV would not do much except document the craft's condition, both for historical and operational reasons. I certainly wanted to have some clean footage of the spacecraft on the bottom before we messed the area up with the vehicle, which was bound to happen. Possibly on that same dive, we'd remove the dye marker canister which should be resting nearby. I had never been able to learn how it was actually attached to the capsule, but after almost 40 years on the bottom, it should come off pretty easily. Dive number two would require attaching two or three recovery tools to the top of Liberty Bell 7. In 1992, before we first looked for Liberty Bell 7 using the DOSS side scan sonar, I got Oceaneering to manufacture four special clamp style tools designed to fit around the capsule's escape tower mating ring.

The area where Liberty Bell 7's escape tower mounted was hardened. However, even after talking to Dr. Max Faget, the capsule's designer, I still didn't know what the ring was made of; this would significantly determine whether or not it would support the capsule's weight. I figured it was either titanium or stainless steel, as it was non ferrous (i.e., non magnetic); a fact I determined after studying Mercury spacecraft wreckage from the failed Mercury-Atlas No 1 mission at the Kansas Cosmosphere and Space Center (Max Ary had managed to acquire the wreckage from its owner, who purchased it at a government auction). To me, the ring, which had a cross section similar to an "I" beam, *looked* like it was stainless steel. The four recovery tools used small metal fingers to clamp around the inside and outside of the top of this ring; each one of these fingers was machined with an interior and exterior radius specifically for this purpose. However, after the failed 1993 expedition, all but one of the tools vanished (though I kept one in my closet for several years).

When at least two (three would be better) of the tools were attached to the capsule's top, we'd be able to connect a nylon strap to the tools giving us a single lift point that could be connected to a kevlar recovery line. Given the water depth, I estimated that we'd need at least 17,000 feet of line, the more the better. We could store the line on the deck of the recovery ship and haul it in using a traction winch. Or even better, we could simply lift the capsule using the Magellan, which had a through-the-frame lift capability of over 2,000 lbs. The spacecraft could be transferred to another line once we got the ROV near the surface. One way or the other, I knew the Magellan could do the job, *if* we found the capsule. I had answered my last question. Liberty Bell 7 could be found, was worth recovering, and could be raised.

It was about this time that Jim Banke, a reporter for Florida Today decided to do an article on Liberty Bell 7 and the idea of recovering the capsule.

He called me and I gave him some photos and information about what I hoped to do. Normally, I avoided publicizing what I was working on, but I figured that at this point in time I had nothing to lose. It was a Florida paper and I doubted anyone at Oceaneering would see the article.

Right after the Florida Today piece came out I got a call from Jim Ball, NASA's general manager of the Kennedy Space Center Visitor Center. He told me that he was real interested in the idea of recovering Liberty Bell 7 and putting it on display in a new building they planned to construct. The whole thing sounded perfect.

For once, I thought I had all the pieces in place: A group who had a reason for wanting the capsule (to display it) and the money to pay for it. I started working on a proposal for the visitor center and the Delaware North Corporation, the company that ran the facility for NASA and the ones who would have to come up with the funding. They were a tough sell since I didn't think they were too enthusiastic about risking over a million dollars to find and raise a 1961 space artifact.

Myself and John Bouvier, a friend who was helping me out with the business aspects of the project, both flew down to Florida to check out the visitor center and Delaware North. It was starting to look pretty good. I really thought something would come of these efforts. Then nothing happened. After more conversations, once again, and at our own expense, we flew down to Florida for a meeting, where we were supposed to start contract negotiations. However, all we did was talk about the project some more and get nowhere. It was a waste of time and money. Slowly but surely, the KSC Visitor Center and Delaware North moved out of the picture. From what I later found out, Delaware North thought the project was too risky. They made a serious mistake.

After the fiasco in Florida, I essentially tried to put the whole idea of finding Liberty Bell 7 out of my mind. I mean, after over ten years of work, what had it gotten me? Nothing but headaches and a lot of bills. I had neglected my career and my life for too long. I decided that it wasn't worth it. I was paying too big a price. I gave up.

I kept working on the Sea Tractor and started doing cable work on the US Navy's USNS Zeus (operated by the Military Sealift Command), their largest cable ship, with their CRS 1 and 2 underwater vehicles. It was a good job as the Zeus was a very comfortable ship and for once, we had all the spare parts and tools we needed. I also started baby-sitting some of the Navy's smaller ROVs at our facility in Maryland. Even though the little Sea Rover vehicles were not as impressive as the larger systems like Deep Drone and CURV III, I was beyond impressing anyone. What was important was that the smaller vehicles were more reliable and easier to mobilize. It was a good job. Overall, by 1998, things were looking up for me. I then heard that Oceaneering might be doing some dives on the Titanic for the Discovery Channel and that they were putting together a crew. I lobbied hard with one of my bosses, Chris Klentzman, to go on the operation. Mostly, because diving on the Titanic was one of my lifelong dreams and I knew I could do a good job driving the Magellan. Fortunately, Chris was sympathetic and worked to get me involved. I figured that even if I never found Liberty Bell 7, at least I will have completed one of my goals in life. I managed to get added to the Magellan's Titanic team, knowing that after spending three weeks at sea on that job, I would have only three days at home before flying to England to do 45 straight days of work on the Zeus. I didn't care. I wanted to go.

By late July in 1998, I was working on the Research Vessel Ocean Discovery, chartered by the Discovery Channel to serve as a platform for both the Magellan ROV systems and NBC Dateline's production studio and video uplink facility. I was more than impressed with the NBC technicians and we made it a point to loan each other parts and consumables. I don't think any of us minded the long hours in the hot sun, knowing that if everything went as planned, we'd soon be diving on the most famous shipwreck in the world.

Sometime before I left for Boston and the Titanic job, I helped a local Washington film producer and director, Joe Casey, compose a letter to the Discovery Channel telling them about the work I had done over the years on the Liberty Bell 7 project. Joe had been talking to Abby Greensfelder, one of Discovery's project development coordinators, about the project and thought they might be interested in producing a short documentary film about the flight, as well as the possibility of finding the spacecraft. I told him that Discovery had rejected the project years ago and it was a waste of time. However, I did have a lot of stock footage from the 1993 expedition and felt that maybe they could make it into a film that would help promote the idea. Maybe for once I could actually get paid for something. I wrote a draft letter for Joe and forgot about it, figuring it was all the usual waste of time.

I was on board the Ocean Discovery, hooking up some electrical cables when George Brotchi, Oceaneering's hard-nosed project manager, got my attention. "You've got a phone call," he said. Some guy from the Discovery Channel, who I didn't know (it was Bob Sitrick) handed me his cell phone.

Space Money: One of the Silver Certificates discovered wrapped around a wiring bundle inside of Liberty Bell 7. This particular bill (five were found) was signed by some of the launch crew. KCSC Photo.

Orbital Globe: Liberty Bell 7's Earth Path Indicator globe survived nicely in the freezing waters of the deep ocean. The component is a plastic sphere covered with an overlay of the Earth's continents. KCSC Photo.

Igniter: The knurled cap for the explosive hatch igniter was found under Grissom's couch, nearby the Randall survival knife. The steel cotter pin which safed the plunger was never found. KCSC Photo.

Left Console: This section of Liberty Bell 7's left console contained some of the controls for the Automatic Stabilization Control System (ASCS). The red "DECOMP" Handle on the left, when pulled out away from the control panel, vented the capsule's atmosphere to space. Pulling the white "RECOMP" handle to the right filled the cabin interior back up with pure oxygen. KCSC Photo.

Signage: The René 41 panel displaying the name "LIBERTY BELL 7" shows well after cleaning. As with all of these exterior panels, the over-sized mounting holes allowed for expansion of the metal during reentry heating. KCSC Photo.

"Who?... me?" I couldn't imagine who was calling me on a cellular phone in Boston. At the time, we had about four diesel-powered welding machines roaring full blast on the back deck and the noise was deafening, to say the least.

As I walked forward towards the Magellan control van, I screamed into the phone, "Hang on a minute until I get someplace quiet!" Inside the van, it was quieter, but still pretty noisy.

"Is this Curt Newport?" the female voice on the phone said to me.

"Yeah."

"This is Tom Caliandro's assistant at the Discovery Channel and Mr. Caliandro wants to talk to you." What in the world could this guy I didn't know want from me? Had I done something wrong? I was clueless, to say the least.

"Curt?" the telephone voice said.

"Yes, what can I do for you?"

"We wanted to talk to you about Liberty Bell 7... To learn more about the project..." I was flabbergasted. Discovery? Interested in Liberty Bell 7? I still had their rejection letter from 1991:

> While we were impressed with both the concept and the treatment of *The Search and Recovery of the Liberty Bell 7*, we regret that this project does not meet our production needs at this time.
> — Michelle Kaplan, The Discovery Channel, February 4, 1991

Back then, I figured that Discovery was more interested in showing films of dolphins and other wildlife, as opposed to deep ocean salvage operations. Besides, who in the hell was going to pay for it all. Discovery? No way. However, there was a lot I didn't know. The Discovery Channel had made a lot of money in the eight years since I got that letter. In addition, Mike Quattrone, their Vice President, has just initiated a new program called "Expedition Adventure," where they sponsored their own field expeditions and broadcast the related documentary films. Another thing I didn't know was that, unlike everyone else I had been talking to over the years, Mike Quattrone *liked taking risks*.

Caliandro basically grilled me for about 20 minutes on all aspects of the project: When could it be done... how much might it cost... what type of equipment would be needed... and how Liberty Bell 7 could be raised. Of course, by then I could recite it all in my sleep and rattled off answers to every one of his questions. I had done so much work on Liberty Bell 7 that there were things I was *never* going to forget. We agreed to talk again the next day and do a teleconference with other Discovery types.

As I walked out of the van to get back to work, Brotchi, wearing his trademark tee shirt and suspenders commented, "You must be an important man!" I was still in a little bit of a daze and grumbled some sort of response.

The next afternoon, with me talking through a pay phone near the dock, several Discovery Channel executives and I kept talking about the project. As usual, I had answers to all of their questions: . . . we would need a deep ocean side scan sonar system and ROV to locate and recover the capsule. . . would probably take one to two weeks to search the area. . . could mobilize from Port Canaveral, Florida. . . we could do it next spring, before the hurricane season. . . and it'd cost at least a million dollars.

What I found amusing at the time, was that they seemed to be worried that I might go with someone else on the project, as though the idea of recovering Liberty Bell 7 was some hot property that other television networks were going to try and steal. I didn't tell them that everybody, and I mean *everybody*, had long since blown me off with respect to Liberty Bell 7. If Discovery passed (which I figured they would as soon as they found out what it was really going to cost), I had no idea who else to go to because I wasn't even really working on it any more. I had given up, right? They faxed me a letter of agreement which I studied very closely. While I would have liked to have an attorney check it out, that was impossible at the time and after they clarified some of the wording for me, I signed it, faxed it to Discovery, and tried to concentrate on the Titanic. I was just going through the drill, assuming that like so many times before, nothing was going to come of it.

The Ocean Discovery departed Boston for the Titanic's wreck site as we continued getting the Magellan ready to dive. I even scrounged some circuit breakers from the ship's spares for our new underwater lights. Our job was to transmit broadcast quality video from the wreck to the surface, where the NBC people would shoot it up to a satellite and hopefully into several million households around the world. To do this, we integrated a special camera system developed by the Woods Hole Oceanographic Institution which consisted of high resolution and high definition cameras, each of them mounted in slick titanium housings. The Magellan also had four 450 watt HMI lights mounted on a light boom which would illuminate a large area around the vehicle.

I was piloting the Magellan when we finally sighted the Titanic's bow section, and at the time I had little more than an hour of driving time on the Magellan. Fortunately, it was an easy vehicle to drive, if you knew what to do. Driving the Magellan is all about power management in that you have to recognize that the vehicle is only fitted with a 25 h.p. hydraulic power unit. When trying to position the vehicle relative to any object, it is impossible to get any useful thrust out of the vertical and axial propulsion units if you try to use them at the same time. You have to work with the vehicle to get it to do what you want. I was very impressed with the Magellan because, after a complete rebuild, it made it all the way to the bottom, almost 13,000 feet down, on the first dive and stayed at depth for over nine hours. With our 1,800 watts of light, I could float the Magellan off the Titanic's starboard rail and see the port side of the ship. Amazing.

I think of the Titanic as the ultimate ROV playground. All of the cables which might entangle an underwater vehicle have long since been removed and you'd have a hard time getting a vehicle stuck on the wreck, unless you tried exploring the interior. I could start at the ship's starboard rail, cut over to the fallen forward mast, drive up that, circle around the bridge, and then scoot down the deck while keeping the vehicle just inches off the ship. I was not afraid to get real close to the sides of the ship and normally used the Magellan's sonar to gauge the distance between our light booms and the Titanic's hull.

Soon after we first sighted the wreck, the NBC television producers asked me to drive the Magellan from the bottom in front of the ship, right up the knife edge of the bow so they could get their "dramatic" shot. On the first try, they said, "you're too far away." I got closer. On the second try, they complained, "you're going too slow." Finally, I said to hell with it. They want close and fast? I stuck the front of the Woods Hole camera maybe a couple of feet from the bow on the bottom and started up, pretty fast. Unfortunately, I forgot two things. First, that the Titanic's bow has a little rake to it. Second, that the ship's bow section is actually bent down, raking the front edge even more. As the Magellan zoomed up the bow, I held it right on center, until the camera's lens started knocking off pieces of "rusticles," hanging down from a large steel shackle attached to the bow. I know I kissed the Titanic's bow with the high resolution camera because when the vehicle was recovered, the camera had a few smudges on its lens. I was later reminded that the camera's "optically perfect" lens cost about $40,000 to make. I was lucky I didn't destroy it.

The Magellan wasn't the only underwater vehicle diving on the wreck during the 1998 expedition. IFREMER, the famed French oceanographic firm, had their Nautile manned submersible on board the RV Nadir, their research ship. They also had a 5,000 meter capable ROV on a chartered supply boat with which they hoped to dive on the bow section. However, the vehicle never made it to the bottom and they experienced numerous technical problems. While I had no way of knowing what type of difficulties they were having, I felt that they gave up too easily. Ron Schmidt, our Project Manager for the Magellan, even offered to transfer to their ship and help them sort things out. They refused. The French were (and are) highly territorial.

After two weeks of operations on the Titanic, we headed to St. John's, Newfoundland, to disembark some personnel, myself included. I had told Discovery earlier that I would call them when I got off the Titanic job so we could set up a meeting. I got home on a Wednesday, scheduled a meeting for that Friday, and got ready to leave for England that Sunday. The meeting, which was with several higher up Discovery executives went well and as usual, I had a hard time remembering names (it's always been a problem for me). When we got done, knowing that I had to fly out of the country in two days for as much as a month, I said, "I guess you want a proposal, right?" Of course, they needed a proposal and there was no way I was going to wait until I got back from England to give them one; gotta strike when the iron's hot. I promised them that they would have something on Monday. Before I left, Mike Prettyman, their business manager, asked me, "How do we get in touch with you while you're gone?"

"You don't," I replied. "We're going to be doing classified operations and the ship is under radio silence when we're at sea." That raised a few eyebrows and they didn't seem to like that.

However, I added, "If we get into port during the operation I'll give you a call."

The last thing I did at Discovery was meet the man who might be signing the checks, Mike Quattrone, the Vice President and General Manager of the Discovery Channel. We chatted for about five minutes talking in general about the possibilities of raising the capsule. Mike also mentioned in passing that Discovery had been approached by three other groups about Liberty Bell 7, but "didn't think they had their act together." Other groups? I wondered who they could be.

Before I left, I nervously looked Quattrone right in the eye and said one last thing: "If you want to find and recover Liberty Bell 7, you are not going to find anyone who knows more about it than me." I meant every word of that truth. In the 48 hours before I left the country, I finished the draft of a technical proposal for the Liberty Bell 7 project, drawing on all of the writings I had done on the subject over the past decade. Considering how little time I had to create it, it was a fine document. By the time it was delivered to Discovery, I was already in the south of England, getting ready to go to sea on the USNS Zeus, and wondering what was going to come of it all.

There was one thing that continued to bother me. It was something I had read years earlier in the capsule's Configuration Specification document; something about the pyrotechnics on the spacecraft:

> *Two SOFAR bombs shall be installed. In a normal landing sequence, one of these (2,500 feet sound ranging) shall be ejected at main chute deployment. The second bomb shall be used to transmit sound ranging signals at a depth of 3,000 feet and shall be permanently mounted to capsule structure.*
> — MERCURY CAPSULE NO. 11 CONFIGURATION SPECIFICATION, MERCURY REDSTONE NO. 4, McDonnell Aircraft Corporation, March 6, 1961.

Did Liberty Bell 7 have a bomb onboard?

Chapter 7 – Target Number 71
Liberty Bell 7 Discovered

Fortune's a right whore. If she give ought, she deals it in small parcels, that she may take away all at one swoop.
— John Webster

From the time I returned from England in early 1999, Discovery and I continued talking about Liberty Bell 7 as I fine-tuned the expedition budget and worked to alleviate any of their concerns. As incredible as it sounds, even with all that I had taught myself about Liberty Bell 7, I had a few doubts about the project. It wasn't that it couldn't be done. It was just that there are many more things involved than simple theoretical capabilities. Even if you had the right equipment and personnel, there were loads of things that could go wrong and cause you to fail in the end. As with any undersea operation, the key was to be able to accomplish your objective while spending a fixed sum of money using fallible equipment and personnel in the unpredictable ocean environment. Not an easy task. Imagine being NASA and having a Saturn V Moon rocket sitting on the launch pad, ready to go. Theoretically, if every component of the booster worked as designed, the rocket should lift off the pad and carry the command module into orbit. But what about the weather? What if it got bad enough to violate launch protocol? The launch would have to be postponed, causing some serious cost overruns. What about if something failed on the rocket after launch? The whole effort would have to be aborted as the Saturn V crashed into the Atlantic Ocean. Yes, in theory, it was possible to find and recover Liberty Bell 7. However, would everything work as designed? Would the weather stay good enough to allow the operation to continue working? Or, would we have to "abort" and hope that we had enough money to carry on?

I knew what elements we needed to have a realistic shot at finding and recovering the capsule; the problem was getting them. Every time I submitted my budget to Discovery, it was too expensive and after a while I felt as though they were determined to make it impossible to succeed. I mean, if we were going to go through all the trouble and expense to attempt the expedition, it should be an effort where we had a realistic chance of success. As a result, when calculating the costs, I did what anyone in the business does: try to hedge their bets by incorporating some contingencies. I felt reasonably confident that the capsule was inside of my basic 24 square mile search area. But was I positive? No, because there were two many unknowns. I knew what the *theoretical* accuracy was of NASA's FPS-16 tracking radars. However, there was no way I could be sure what that meant in terms of trying to locate a sunken object lost in 1961. I still didn't know for sure what the subsurface currents were like because few people had ever done any work in the area where Liberty Bell 7 sank. Given that I estimated that the spacecraft took as much as 1 to 1½ hours to hit bottom, it could have drifted quite a ways in any significant current. As a result, my original search area was doubled: 48 square miles. I *really* thought the capsule should be somewhere in that box. However, as before, the cost was too high. What was I supposed to do? I could put my foot down and say, "Sorry, but this is simply what it's going to cost." However, maybe Discovery wouldn't agree to sponsor the project. Was I willing to throw away what might be my only real chance? I couldn't do that. Maybe all the gear would work OK. . . maybe the weather would stay good enough to work. . . hopefully, nothing would break . . . I knew that I was really kidding myself. I had worked at sea long enough to know that nothing ever went exactly as planned and there were many things that could go wrong. Discovery started talking about getting "expedition insurance," where a documentary film project can be underwritten if the weather gets bad and they can't film; I told them that if they wanted insurance, charter a larger ship. You can't get better insurance than that.

When all was said and done we still had good equipment and crew. However, we only had 13 days on location to search my downsized 24 square mile search area, evaluate all of the sonar targets (there couldn't be much else there, right?), find Liberty Bell 7, rig the capsule for recovery, and lift it to the surface. If we had any major technical problems with the gear or got nailed by some bad weather, we were out of luck.

Discovery did not actually agree to move forward on the Liberty Bell 7 project until late 1998. I was still working in my normal capacity at Oceaneering, as an Assistant Project Manager in charge of three of the Navy's Sea Rover ROVs. Once I knew that we were definitely going to do the project, I asked Chris Klentzman, the man in overall charge of the Navy contract I was working on, for a leave of absence. Chris was as supportive as he could be and we managed to hash out an acceptable agreement. I could have the time away from work as long as I used Oceaneering personnel and equipment on the job. That was fine with me as I thought they were the best ones for the operation anyway.

What followed was a blur of activity; three months of the most intense work I have ever done. First, it took $11,000 in legal fees to reach a reasonable agreement with Discovery on my role in the project, money I could have definitely used for other things. While I had a good attorney, to this day lawyers and contract negotiations make me want to vomit. I was drinking far too much (from the stress), getting very little rest, and still having to go back and forth with Discovery every day from morning to night. On the final day of negotiations, I accidentally swallowed two Melatonin tablets instead of aspirin and finally caved in. I was walking death as I chain smoked, drank bourbon, and tried to keep everything on track. I decided that the important thing was to try and find Liberty Bell 7, not to squabble over legal issues. In other words, if I held out over some stupid contract wording, I would stay home. If we finalized a contract, I could go to sea and look for Liberty Bell 7. I decided to go for it. I felt like Rocky Balboa finally getting a chance to play with the big boys. Weren't dreams called that because they never actually happened? How could it be that I would get a chance to actually do what I had been talking about for over a decade?

It was like building a house of cards. The cards were the numerous elements that made up the project and they were so very fragile. All it would take was one good breath of wind and the whole thing would come toppling down. I also convinced NASA to sign off on an agreement whereby they would pass title to Liberty Bell 7 from NASA to the Discovery Channel. When the spacecraft hit the deck, the Discovery Channel would be the proud owner of a 1961 Mercury spacecraft, serial number 011, only used once. When the craft came off the ship and landed on the back of a flatbed truck, it became the property of the Kansas Cosmosphere and Space Center, a soon-to-be affiliate of the Smithsonian Institution. The Cosmosphere had agreed sight unseen to disassemble, clean, and reassemble Liberty Bell 7 into displayable condition in return for eventual ownership of the spacecraft.

A company called UXB International was retained to locate and dispose of one SOFAR bomb (a device designed to mark the location of a sunken Mercury capsule) that was supposed to be mounted somewhere inside of Liberty Bell 7. I didn't know for certain if the bomb had exploded or not. According to NASA's Post Flight Memorandum, the SOFAR listening network never detected the detonation. But did that mean it didn't explode, or did they just not hear it for some reason? If it did in fact explode, there would be a lot of damage to the capsule's structure. If it didn't, the UXB technicians would deal with it.

I also found someone at NASA to "unofficially" review all of my research. While I felt that I had gained a good understanding of the technical aspects of Grissom's mission, I really wanted someone else to check my work. I was not an expert in trajectory analysis. However, Ray Silvestri at the Johnson Space Center was. Max Ary, the president of the Kansas Cosmosphere and Space Center, earlier called Gene Kranz, the famed NASA Flight Director, on my behalf, whereby Kranz said, "Ray Silvestri is the guy you want." I sent Silvestri much of the trajectory data I had collected, along with my interpretation of what I thought it all meant. He did some rough calculations on the capsule's suborbital mission and assessed the flight data using his past experience with ballistic trajectories. During our last conversation, we compared notes via a cellular telephone on the Port Canaveral dock just before sailing. Ray generally confirmed what I thought I knew: in theory, Liberty Bell 7 should be somewhere within my primary search area.

Oceaneering hired the John Chance Company to do all of the navigation on the expedition and over a period of weeks, I furnished them with all of my historical navigational data. Once they had plotted the locations out on a large chart, I could have a look at everything and finalize the search area. One thing that many people don't realize is that the shape of the Earth was defined differently in 1961. At that time, all of NASA's navigational data was based on what was called the Clark Ellipsoid of 1886. It is one definition of the shape of the planet. However, today, navigators normally used the WGS-84 Ellipsoid (World Geodetic System for 1984). The difference between the two was only about 300 meters, but I figured it would pay to be as close as possible: when we searched for Liberty Bell 7 we would be navigating using the same Ellipsoid NASA used in 1961 (ironically, on the second Liberty Bell 7 expedition, our new navigator would be unaware of this fact and unwittingly use the standard WGS-84 Ellipsoid, causing us to waste over two days trying to relocate the spacecraft).

Instead of shuttling to Oceaneering and Upper Marlboro, Maryland every day, my days became 18 hour affairs of faxes, emails, conference calls, and typing; they say be careful what you wish for and in my case it was true in spades. I still found it hard to believe we were really going to try it. In July of 1998, Liberty Bell 7 was what it had always been: an underwater pipe dream. Now, little more than six months later, I had a contract with the Discovery Channel, a side scan sonar, an ROV, and a crew of people getting ready to go to sea to try and find a 1961 spacecraft lost at sea for almost 40 years.

By early April, I had moved down to Cocoa Beach, Florida, and was camped out in a hotel room in Cape Canaveral. My mind was focused on one thing and one thing only: making sure that I did everything possible to enable us to locate and recover Liberty Bell 7. Anything that was not related to that one goal, I ignored.

Even today, Cocoa Beach is a place rich in space lore, from the Holiday Inn where the Mercury astronauts exchanged practical jokes, the hotel across the street with a swimming pool emblazoned with the Mercury 7 insignia, and the famous highway A1A, where the astronauts raced each other in their hot corvettes. Located on Florida's east coast about half-way down the state, it doesn't get too cold in the winter and the area lacks the fatiguing summertime humidity of the Ft. Lauderdale area. Overall, it's probably a good place to live and work. These days, everything in Cocoa Beach is identified relative to the massive Ron John's Surf Shop and the locals are still part of the space program, many of them making regular trips to the coast to watch the Shuttle launches.

It was not long before Oceaneering and our support ship, the Needham Tide, both arrived in Port Canaveral. As I drove into the commercial docks for my first look at the ship, I saw the Magellan ROV and related gear sitting strapped to a flat bed truck on the dusty gravel road to the pier. The ship was nothing impressive, simply an average run of the mill offshore supply boat that was a few years older than normal. The Tide had a bow thruster for docking, two fixed pitch propellers, and a Gulf of Mexico crew. I checked out of the hotel on April 18th and moved on board, anxious to get to sea.

When Dr. Robert Ballad discovered the remains of the famed White Star ocean liner, the Titanic, he had one of the most advanced and maneuverable research ships in the world, the RV Knorr, with its cycloidial propulsion system. With that highly-sophisticated rotating assembly of propellers, the ship could spin on a dime and move at very slow speeds in any direction.

In addition, the target he was searching for was the Titanic, 882 feet of steel ocean liner, which was broken into two large sections (one of them almost 500 feet in length), and sitting amongst a mile long debris field. Some of the ship's massive wreckage in the debris field was over 40 feet tall. Also, the bottom terrain in the area was generally flat with few abnormalities. Plus, it was "only" 12,600 feet deep.

My crew and I were sailing with an aging 180 foot Gulf of Mexico "mud" boat, called that because it is used to haul drilling mud to oil rigs. Because the

Tide had fixed pitch propellers, it was incapable of moving at less than four knots even with only one screw engaged. Our target speed for the search phase was 1.5 knots along our track lines.

And then, of course, there was our target. Tiny by any standard, you could fit several Mercury capsules into only one of the boilers from the Titanic. In fact, the spacecraft was smaller than one propeller blade from that ship. Even worse was the fact that we had to locate and identify this tiny object with no massive debris field to guide us. One little heave of the sonar at the wrong time and Liberty Bell 7 would elude our sonar.

Then there was the terrain. During my previous experience in the area, known as the Blake Basin, I discovered to my horror that the location was characterized by numerous "sand waves." The crests of some of these sand dunes approached 50 feet! Definitely tall enough to hide a nine foot tall spacecraft from our towed sonar. Finally, there was the water depth. Liberty Bell 7 was thought to be in about 15,600 feet of water, a half mile deeper than the Titanic. It would take us anywhere from six to 10 hours just to reposition the ship after each search line. Even if we managed to image the capsule, Liberty Bell 7 would look like little more than a tiny trail of bright pixels on our computerized display. To put it all into perspective, you'd have to stack 28 Washington Monuments on top of each other to equal the water depth. Simply getting our underwater vehicle to the bottom would take four hours. When towing our sonar at depth, if our ship was over Arlington National Cemetery, the sonar would be somewhere around the United States Capitol, over three miles away.

What we were attempting was a "technological challenge" in spades. In fact, in some ways finding Liberty Bell 7 would be blind luck. Still, I tried to minimize our chances of failure by making sure that the personnel for the venture were top-notch. For the most part, the operations team supplied by Oceaneering International was a mishmash of people from both government and commercial operations. Some I knew and others I had never met. But the two key people I had asked for were among the crew: Ron Schmidt and Mark Wilson.

Ron was a mechanical engineer who had burned some bridges with Oceaneering along the way by leaving their commercial division to work with a new US Navy underwater vehicle. While I wanted him to be Oceaneering's Project Manager, Ron ended up running the night shift. Ron didn't speak much, but when he did, he was worth listening to. He was a superb ROV pilot and instrumental in building the Magellan 725 vehicle which we planned to use to recover Liberty Bell 7. He was hard to know but easy to have confidence in.

Mark Wilson was a right wing computer expert of sorts. But like Ron, he was a gifted electronics technician who had considerable experience with both the Ocean Explorer sonar and Magellan 725 ROV. He knew the Ocean Explorer sonar like no one else. Mark could scrape salt water corrosion off a circuit board and breath life back into it; a skill that would certainly be put to the test during our search for Liberty Bell 7.

Steve Wright, the Project Manager supplied by Oceaneering, was a former diver who was working as an engineer in the company's advanced technologies division in Maryland. While I knew him and we had always gotten along before, our relationship on the Liberty Bell 7 operation became strained over time, mostly due to disagreements over what constituted a working side scan sonar.

We had numerous other people on board the Needham Tide. Will Handly from the Woods Hole Oceanographic Institute would maintain the special cameras on board the Magellan 725 ROV, the same ones we had used to image the Titanic the previous year. Peter Schnall along with his film crew from Partisan Pictures would document the events of the operation for the Discovery Channel, and a couple of navigators from the John E. Chance company would make sure we were where we were supposed to be.

When I moved on board, Oceaneering was still outfitting the sophisticated cameras supplied by Woods Hole and trying to get our launch and recovery crane working as designed.

After a one day delay waiting for crane parts, the Needham Tide headed for open ocean just after noon on April 19th. As I leaned against the port rail I could just make out launch pad number 5 where Grissom was thrust into the heavens almost 40 years before. The ancient Redstone booster now sitting on the pad made it look like a church steeple. I imagined the ghost of Gus Grissom in a silver pressure suit mocking us from the pad; shaking his fist at us. Was that gesture a threat or encouragement?

I also began to think about the expedition ahead. This would be my third trip to the Liberty Bell 7 site and I wondered how it would turn out. My last chance? Probably. If we failed this time I could see no logic in pursuing my objective further. However, I still had something to prove. While I felt I was respected in the subsea business, I was not known as an organizer or leader of such expeditions. I wanted to show people that I could assemble a team and run a successful mission.

Mark Wilson and I stared at the navigation screen as the Needham Tide wallowed in the swells, her throbbing engines protesting like bucking horses as the helmsman engaged one of the ship's screws. This was a critical test, one that would see if our ship could steam slow enough ahead to tow the sonar.

"Four knots!" I exclaimed. "That's as slow as they can get on one engine?"

Mark grimaced and thumbed the microphone, "Bridge. . . sonar. . . you on one screw now?"

"Yep. . . that's it on the starboard engine."

I put my two palms against my head and squeezed like a vice. "They'll have to do better than that, that's for sure," I said.

The ship driver on the bridge eased the starboard stick a little forward and put the port engine in reverse. More vibration filled the ship. The high pitched whine of an air starter added to the rumbling sounds. Even the bow thruster was now running.

The ship driver's tired voice crackled from the speaker, "How's that look? I got the port one astern now." The rumbling sound rose, making the ship shake like a herd of cattle galloping across our deck. Coffee in a nearby Styrofoam cup danced in unison with the vibration.

The numbers fell slowly. . . 3.8. . . 2.5. . . 1.9. . . 1.4. Scratching my head with nervousness, I gestured at the screen to Mark. "What do you think? This going to work?"

"It should, as long as they don't mind doing that 24 hours a day for a week." We were in business. Running one propeller in reverse and the other one forward, we had created our own artificial variable pitch propulsion system.

The engines were reengaged and the Needham Tide chugged eastward, towards the location of America's only sunken spacecraft. But before we could start the actual search, we needed to do a test dive with the Ocean Explorer 6000 to confirm that our Louisiana mud boat could drive a straight line with the sonar in the water, and also to make sure the Ocean Explorer worked at depth.

We throttled-down once again.

The ship started vibrating and rumbling as the helmsman tweaked one throttle ahead while easing the other astern. The ocean surface behind the port propeller churned as the bronze blades clawed the sea, slowing the ship down to a respectable pace.

Oceaneering's team strapped on life vests, plopped on hard hats and lit up cigarettes, as Richard Dailey took his seat on the Magellan's launch and recovery crane. Richard was nicknamed R.D. Although I had never worked with him before now, he was easy to work with. He was also rumored to be very lucky, a fact supported by his role in several previous underwater operations that somehow managed to succeed in spite of tremendous adversity (later on, I heard the exact opposite; that R.D. was very unlucky).

Getting the Ocean Explorer sonar in and out of the water was tricky business. It was made up of two separate components, the depressor (a large finned tube of steel filled with lead) and the search vehicle, a long yellow torpedo-shaped device fitted with floatation, lead ballast, electronics bottles, cabling, and four acoustic transducers. To get it into the water, the search vehicle was deployed first on the end of about 300 feet of flexible tether. As the ship moved forward, the depressor was then launched from the same crane, dangling from the end of several miles of rigid triple armored steel tow cable, as the whole mess was lowered into the deep sea.

Diesel engines screamed and belched black smoke as R.D. was given the signal to hoist up the depressor. Several hundred pounds of lead slapped the bottom of a kick plate while the cable reel and traction winch tugged hard on the armored umbilical, locking the mass in position. Deck hands grasped the rails on the sides of the sonar as it lurched off the deck and into the air, held aloft by R.D.'s crane. As the Needham Tide crept slowly ahead, R.D. swung the crane until the end was pointed directly aft.

"OK... lower it down," Steve screamed as the lift line dumped the sonar into the sea. The noise was deafening as massive hydraulic motors strained against thousands of pounds of weight and the ship heaved gently in the light ocean swells. The depressor, looking like a small bomb that had been dropped numerous times, followed the sonar as steel cable was paid out, dragging the sonar beneath the seas towards the ocean floor.

Inside the sonar van, Mark intently monitored the computer displays looking for any sign of trouble. It came in about ninety minutes when wiring in the soft tether shorted out.

After the sonar was back on deck, Mark discovered damaged wiring in the tether and spent an hour repairing a splice in the cable. The sonar was again launched.

This time it lasted about an hour before the same problem forced us to bring the sonar back on deck. This time, we decided to replace the entire cable.

Soon the sonar was in the water as Mark slowly rotated the knobs in the sonar van, coaxing as much performance as possible out of the equipment. Our ship cooperated by moving ahead at about two knots. We left the sonar in the water until we were convinced that everything worked. Then we recovered the sonar and chained it to the deck, the vehicle still dripping with warm salt water. We were ready now and headed toward Liberty Bell 7.

Our ship sliced through the rolling swells at full speed, directly underneath Grissom's 1961 flight path. The morning sun crept over the horizon, filling the deck of our ship with warmth as the helmsman slowly eased off the throttles. The sea was nearly a flat calm as our eyes adjusted to the blinding light, all of us still sleepy but excited. Even though the job had yet to really start, we were all starting to feel the tension. Coffee flowed down our throats, cigarette smoke filled our lungs, and steel toed work boots were laced up tight for what was to be a long workday. In the pantry, Emily, our cook, fried up bacon and eggs as we all shared our anticipation at what we would find during the search.

Based on my historical research and analysis of the flight path, I had previously established our search area as an elongated rectangle eight miles long and three miles wide. This was aligned along the space craft's flight path as well as the direction the capsule drifted by parachute before landing in the ocean. We would initially make several passes through the area, each 1,000 meter swath slightly overlapping the next. There was a big problem, however. The Ocean Explorer had a blind spot directly underneath the sonar. Consequently, the sonar could pass directly over a dozen Mercury capsules and we would never see them. To counteract this problem, we planned to do several "fill in" lines, where we would tow the sonar in between the previous search lines to image the areas directly underneath the sonar's path over the sea floor.

I had agonized over where to position the search area for months. Due to budget constraints, I had been forced by the Discovery Channel to whack the size of the search area in half. I remembered the discussions during our numerous teleconferences.

"But do we really have to search the entire area? I mean, where do you think it is?" asked Mike Prettyman, Discovery's head of business affairs.

I stammered and steamed a little, "Well, if my numbers are right, it should be in the smaller area, but there are always unknowns with this sort of thing."

"But do you think it's inside the smaller box?"

"It should be, but . . ."

At that time, my head was filled with numbers and angles, locations from Post Flight Memorandums, Firing Test Reports, Trajectory Reports, ship logbooks, preflight locations, post flight and SOFAR bomb locations, as well as surface wind drift data and ocean current projections, all of which had resulted in 17 separate possible locations where Liberty Bell 7 landed in the ocean.

For years, the navigational information had been bouncing around inside of my brain and my head felt ready to burst. Now I was supposed to know where the damned thing was. The problem was that no one knew. It was all an educated guess, one that twisted my gut.

Once again, the diesel engines screamed and spit out smoke as the deck gear was warmed up. Everyone took their places on deck, standing like the front line of a football team waiting for the kick off. In the sonar van, Mark sipped his coffee and flipped his toggle switches, sending the life blood to the Ocean Explorer as it was hauled off the deck and thrust into the ocean, heading to the bottom three miles away. I sat back on a spool of cable and enjoyed the warmth of the sun on my face. This was it. The search had begun. I looked at my watch. It was almost 7:00 in the morning.

After the sonar was in the water and trailing off our stern, Steve stepped in the sonar van and started giving orders.

"OK. . . Give her a little kick ahead. . . try to keep it under two knots."

The steel tow cable angled aft as the ship moved slowly ahead, marking her track by the propeller wake. By 8:43 in the morning the sonar was almost 10,000 feet down, trailing along behind and below the ship like a dog on a very long and costly leash. The ambient pressure on every part of the sonar was just passing 4,454 pounds per square inch. Unbeknownst to us, sea water had started slowly dripping into the vehicle's electronics bottle. Electrical current passed between two of the main power leads. A sensor inside of the bottle tripped and shut the whole thing down.

"Got a ground fault," Mark said.

"Lost telemetry."

Steve cursed, "Damn it. . . all stop on the payout, start bringing it up."

"I'll kill the power to the system." Mark suggested.

"Good idea. Bridge? Sonar. We're recovering; I'll call you when we're ready to bring it out of the water and you can line the ship up into the seas. Just keep going on this same course for now." It was a harbinger of things to come.

Greasy fingers pried off the end of the sonar's electronics bottle as Mark and I peered inside of the long dark tube. Sea water dribbled out of the bottom of the tube, maybe a quart of it.

"Damn."

Salt water had leaked into the bottle, shorting out vital parts of the circuitry. It was now up to Mark to make it right. He dove into the problem as usual, removing wiring and using up several cans of electrical cleaner, all in an effort to remove any salt deposits from the sonar's circuit boards.

It wasn't until 8:00 that night that the Ocean Explorer was ready for launch again. This time the demons of underwater exploration reared their heads much quicker and with far greater fury. At a depth of only 1,000 feet the sonar's delicate innards were fried as sea water once again squirted inside the search vehicle. It could not have been more damaging if a small bomb had gone off inside the electronics bottle.

"Ground fault!" Exclaimed Mark as he watched the needle on the Ground Fault Monitor crash to one side.

"Another one!?" I gasped.

"Sorry, we gotta recover again."

My body went almost limp, burying itself against the carpet lined walls of the sonar van. Another ground fault? What on Earth was going on here? I could not understand how a simple piece of gear could suffer such failures. Side-scan sonars generally work simply because they do not have any moving parts. But in several days, we had not even managed to get the damned thing to the bottom. Failure drifted across the deck like a dark cloud.

Mark extracted the blackened electronics chassis from the Ocean Explorer like a pathologist autopsying a dead man. It did not look good. We were not even sure if the sonar *could* be fixed, never mind how long it would take.

The circuit boards on such systems are custom made, not items than can be purchased off the shelf. If the backplane (the component onto which all of the circuit boards were mounted) was fried, then it could be a show stopper. To replace that component might take weeks, weeks we did not have.

Mark Wilson dissected the intricate electronic components of the sonar like a surgeon, carefully removing each blackened part in turn to gauge whether or not the Ocean Explorer would ever dive again. It was tedious work. Ron Schmidt joined Mark to help in the complicated trouble shooting, the two of them isolated in the sonar van, the rest of us feeling lost, wanting to help but not wanting to interfere in any way. At one point, they reinstalled the electronics chassis back into the sonar, only to carry it back to the sonar van with grim faces and shaking heads.

Many of the electrical connectors on the sonar chassis had carbon deposits on them, all of which had to be cleaned off if we had any hope of continuing the search for Liberty Bell 7. At one point in time, I attached a sign to the outside of the sonar van stating, "No admittance until further notice," to help ensure that they would not be bothered as they toiled over the sonar. Even I rejected the urge to enter the van to ask how they were progressing. I had done enough similar work myself to know when to stay away. This was one of those times.

By Friday, April 23rd we were ready to do a wet check of the sonar to verify that the electronics bottle leaked no more. We dunked the Ocean Explorer to a depth of 1,000 meters and let it sit for an hour. Then we hauled it back to the surface. The interior was dry.

At least that part of the problem was solved, I told myself. But it was still up to our two premier technicians to see if the sonar could be brought back to life.

As we waited for the technicians to repair the sonar, Steve Wright came up with what I though was a crazy idea.

"We need to start searching," he said, puffing on a freshly lit Marlboro.

"Uh, yeah, that would be a good thing. As soon as Mark and Ron are done with the sonar we'll put it back down," I replied.

"No, I mean we have to start searching now; get something done. Oceaneering's losing too much money on this job and if we don't start producing, they're gonna pull the plug."

I took a deep breath and tried to keep from exploding.

"Well we can't search right now, Steve," I said. "The port channel on the sonar still doesn't work right."

Steve adjusted his cap, taking a long drag from a Marlboro. "We can search the way we are."

I got mad. "What the hell are you talking about? You want me to search for this target when only one side of the sonar is working properly? That'll double our search time. Who's going to pay for that?

"All I know is that we've been out here for days, have not accomplished anything, and my boss wants me to get something done."

I stood up and walked away. Who in their right mind would search for something the size of a Mercury spacecraft with a sonar that was half blind? Not me.

Later that night, Ron and I had a private discussion about the sonar and what Steve had suggested. He reassured me that Mark was making progress and the sonar could and would be fixed. Just another 24 hours and we would be ready.

We joked about welding the bottle shut. But the next day, when the Ocean Explorer was finally lowered back into the sea, there were no leaks. We watched the sonar sink below the waves, not to be seen for several days.

It took eight hours for the sonar to hit the Start Of Line, or SOL. Then Mark adjusted its pattern of acoustic energy until it was imaging a 1,000 meter swath of the ocean's bottom. We soon began "mowing the lawn," driving the boat in long lines across the ocean that resembled the rows in a freshly cut lawn. Once that started, we eased into the tedium of the search routine. While we didn't know it at the time, just after the end of the first search line, the sonar detected the hulk of a wooden merchant ship, sunk in the year 1810.

What would the Liberty Bell 7 look like on sonar? Would it be a small conical shaped target? No. Would it look like a rock or any other target we might find? Possibly. The best we could hope for was that Liberty Bell 7 would somehow reflect enough sound to light up a handful of pixels on our dust-coated, finger-smeared computer screen.

On Sunday, the sonar gave up the ghost once more, requiring us to haul in miles of tow cable and once again let Mark the computer wizard tear apart its malfunctioning pile of integrated circuits, resistors, diodes, and power supplies. This time we were lucky. Repairs were quick and once again the towfish was dumped into the deep blue sea.

The search continued. The sonar was dragged over the bottom, spitting out little chirps of low frequency sound which reverberated over bottom silt that had probably remained undisturbed for thousands of years. One by one, we developed a target list – documentation that identified anything we had found that could possibly be the capsule. After three days of dragging the sonar, we had amassed a list of 20 individual sonar contacts. Some of them were undoubtedly dense formations of mud. Others possibly aircraft wreckage like the one I had discovered in 1992 during the first search for the capsule.

By the next day our target list had tripled. We had 60 targets! Where were they coming from? How could so many objects be in such a small area when it was such a huge ocean? We didn't have time to think about that. Before long we had discovered an incredible 88 sonar targets.

All during this time, we were making a valiant attempt to "prioritize" our staggering target list. To even hope to inspect 88 targets was folly. That would take weeks, maybe months. Instead we examined each individual sonar contact, noting their respective strengths and weaknesses, and developing a list of targets of which one was possibly Liberty Bell 7. It was not easy. The information we had available was the recorded sonar data and a stack of target sheets. These were scrawled with notes about each target – the ping number, estimated target strength and general description.

Mark pointed at target number 11 on the video screen.

"I think this is geology. . . it doesn't have much pixel strength, blends in with the terrain. . . see that crest right next to it?" The target was a small smudge that seemed to line up with a ridge a few meters away.

"Yeah. That one's history," I answered.

Mark continued, "OK, now we come to target 12. Not a good candidate; it consists of multiple small targets and we're looking for a single hard contact; I'd cross that one off the list as well."

"OK. Next one?"

That went on for several days until we developed a "hit list" of 16 targets that we liked. These were ones that were isolated, "hard," and did not appear to be part of a geological formation. It was a process of subjective analysis. The sonar could find the targets, but no computer in the world could lead us to the one that was Liberty Bell 7. That would take several human brains, all of them examining the data and making evaluations based on technical knowledge and experience.

On April 29th, 1:07 a.m., the Ocean Explorer was being tugged in the freezing darkness of the deep ocean, three miles below and 1.5 miles behind the Needham Tide. At that instant, the surface electronics of the Ocean Explorer sent a signal down 28,447 feet of triple armored steel coaxial tow cable. This minute voltage was received by the maze of circuitry that Mark Wilson had toiled over for several days. A short burst of electrical energy was sent to the Ocean Explorer's port low frequency transducer and exploded from the side of the sonar, forming a beam of acoustic energy that fanned out at a speed of 4,950 feet per second. In the blink of an eye, it bounced off a hard metal object, returned to the sonar, and shot back up the cable and onto the sonar screen. What registered was a dark colored object protruding about nine feet above the bottom, enthusiastically reflecting the sound that a man-made creation should.

The target was identified as Target number 71, and recorded on our contact sheet as a "hard target across ravine."

We continued our search.

Then another problem reared its head.

Sara Hume, the Discovery Channel producer in charge of this expedition, was giving us only four more days to find Liberty Bell 7. The reason: Money. Like all endeavors, the expedition had a bottom line.

"We really need you to leave the location no later than midnight on Tuesday the 4th," Sarah explained via the crackling connection of a satellite telephone.

I was not surprised, but my still my body nearly folded in agony. "So with or without the capsule, you want us to leave the site at midnight on Tuesday?"

"Yes. I'm sorry, but it's all just starting to cost too much. What do you think your chances are?"

I took a deep breath to relax. "It's hard to say. We're almost done with the search and have tons of targets. We'll just do the best we can. But just so I'm clear on this, on Tuesday at midnight, we're out of here, regardless of whether or not we've found the capsule?"

"I'm afraid that's the way it has to be."

I entered the sonar van. Steve had been trying to convince me to do some cross track lines to look behind the sand dunes. But there was no more time. Frankly, by then I was almost looking forward to the end of the expedition, even one that failed. I was physically and emotionally spent.

"Mark, this is the last search line. When you're done with it, haul up the sonar and switch over to the ROV," I said.

"OK, I'll tell Steve and we'll get it done."

I agonized over the next step. We had filtered over 88 sonar targets down to a manageable list of 16 that looked promising. Now it was doubtful that we could even inspect that number of targets in the time remaining. In the end it was all coming down to money.

I looked at the 16 targets. They were grouped in three areas, which would force us to launch and recover the Magellan ROV three times. That alone would take 24 hours (eight hours per launch and recovery sequence). Even if we managed to locate and visually identify each target in six hours, just finding and documenting each of the 16 targets would take four days. That didn't count the transit time for the ROV from one target to the next. Also, those estimates assumed that the weather remained good enough to dive and the ROV worked flawlessly. Not likely.

Our only hope was to start diving immediately. It was now or never. I immediately began to assess our 16 chances of success, or failure.

Over the course of the search phase, I had manually noted the location of all 88 sonar targets on a large plotting chart on the ship's bridge. In addition, I used several colored highlighter pens and drafting tools to identify each and every target with symbols: a triangle signifying a hard contact, a square for a medium target, and a circle for something I felt was weak.

Looking at the overall picture, it was obvious to me that we should start with a group of six targets clustered around the radar locations generated by the Marshall Space Flight Center (MSFC) and the Langley Research Center's (LaRC) Space Task Group. Though the positions were probably created using the same raw radar data, they were not in the same spot. I chose the MSFC estimate. They had considerable experience with the Redstone booster as well as the ballistic trajectories identical to the one flown by Grissom in 1961. Those six contacts were closest to my best guess as to where Liberty Bell 7 now lay.

It wasn't until 11:00 p.m. on Friday night that I finalized the Magellan's dive plan as Steve joined me on the partially darkened ship's bridge. The two of us huddled over the wrinkled chart, its surface bearing the marks of pencils, erasers, and rulers.

"You see the six targets in this area?" I asked Steve.

"Yeah."

"Let's start with this one, number 71, work our way south towards target 54, then turn east, picking up number 28 until we get to target 61. If we haven't found it by then, we'll head back to north via 14 and finish up this area with target 13."

"OK. What if we don't find it on this dive?"

"Let's worry about that later."

As Steve stomped down the stairs to tell the navigator where we wanted the ship positioned for the morning dive, I could see from the bridge that they were hauling in the last remaining feet of the Ocean Explorer's tow cable.

The deck crane strained under the mass of the sonar as it reeled the dripping cable from the sea. The cable looked like little more than a long silver colored thread, on the end of which was the torpedo-shaped sonar that had finally survived several days deep underwater; enduring forces that would crumple a Styrofoam coffee cup down to the size of a thimble.

Luckily we didn't have to endure those deep ocean forces. But we could not escape the surface forces.

Up until now, the weather had been cooperative. Other than an occasional gust of wind, we had yet to be seriously harassed by the weather. But things were changing. The barometer was beginning to work its way downwards and the long, inviting Atlantic Ocean swells we had grown to appreciate were starting to build ever so slowly.

When we started the predive procedures on the Magellan at 10:26 a.m., on Saturday, May 1st, I recorded in my personal operations log, "wind and seas picking up."

With Ron on deck and Richard Dailey inside of the control van on the control sticks, Dave Warford, the team's resident biker, sucked the 6,000 lb. Remotely Operated Vehicle into the launch crane and started swinging the whole assembly around.

The deck was alive with sounds and motion as the launch crane lurched to a stop directly aft of the Needham Tide.

"ROV. . . deck. . . Everything look OK R.D.?" Ron asked.

"Looks good from in here; still have video and telemetry. . . no problems." Inside the Magellan van, R.D. surveyed the numerous digital readouts, each telling the health of the vehicle.

"OK. . . Dave, put it in the water."

"Lowering away." With that, the three ton deep diving unmanned submersible was dunked into the seas, the long swells partially obscuring it from view.

"Hydraulics on," replied R.D. After flipping a toggle switch on the Magellan's yellowed control panel. The vehicle sprang to life as 2,400 volts of electrical juice flowed into a 25 horsepower electric motor, lighting up the vehicle's hydraulic system.

"R.D. . . spin it around to south," Ron commanded. Inside the van, R.D. twisted the single hand controller or "joystick" fully to the right. On the vehicle, the starboard thruster churned the warm sea water in reverse while the port unit did the same in a forward direction.

As soon as he saw the Magellan pointed directly aft, Ron leaned over a little and gave a long nylon lanyard a sharp tug. The pelican hook connecting the ROV to the launch line popped open and the Magellan was swept away.

The Magellan fought its way off the ship's stern as Ron and a deck hand fed the yellow tether into the sea. After a few minutes, three small plumes of water shot upward from the vehicle's vertical thrusters, driving it towards the bottom, three miles down.

The sub's environment changed rapidly. It was a balmy 75 degrees on the surface. But soon the Magellan would be speeding through the deep, dark abyss. When it reached the ocean's bottom, it would be at a location as desolate and unexplored as the surface of the Moon that Grissom had hoped to visit in person.

Inside the Magellan van, R.D. was joined by Ron as they began the four hour voyage to the bottom.

"Six hundred feet," remarked R.D. to Ron as he fired up the sonar mounted to the top and front of the vehicle.

On the Magellan, the Ametek-Straza sonar started rotating slowly as it filled the area around the ROV with sweeping frequencies of sound that would lead us to target 71.

Two and one half hours later, the Magellan slashed though the 9,000 foot mark. At this point, the pressure on every square inch of the submersible vehicle was equal to the weight of my Saab automobile; about 4,000 pounds. The electrical connectors were being rammed into their sockets by tons of force, and the once balmy surface temperature had plummeted to only 45 degrees. Save for the artificial lighting on the vehicle, the area was totally devoid of any light, except possibly the phosphorescent illumination from nearby sea life.

"Stop payout," Ron said.

"Payout all stopped," answered Dave Warford in between slurps of warm coffee. All of the deck gear, our cable reel, traction winch, and crane sheaves, ground to a halt as we allowed the two miles of steel cable to stretch out. Then we started dropping the ROV again, passing 14,000 feet, well past the depth at which the Titanic lay.

After the Magellan was launched, I retired to my cabin in a vain attempt to get some rest. I knew it was going to be a very long day and had learned before to get sleep when you can while working offshore. I stumbled out of bed in the early afternoon and forced my work boots on. By the time I made my way into the Magellan van, the vehicle was near the bottom.

It was about 2:30 p.m. by the time I took my normal position behind Steve and Mark. Far below us, the Magellan was being driven towards the sea floor by her vertical thrusters. By now, the surrounding water temperature had crashed to a frigid 36 degrees, ambient light was non existent, and the water pressure was an impressive 7,127 pounds of force per square inch.

I quickly realized that all was not well with the Magellan. Noticing that the sonar was not creating any sound, I said, "Mark, what's up with the sonar?"

Mark leaned forward and twisted the gain fully clockwise. "It doesn't look like it's receiving; it appears to be transmitting though." That, we could tell by the little flashes of the amber scan line that were indicative of an acoustic signal. Great.

Peter Schnall and his film crew contributed to the already crowded van as they gleaned the situation from Steve.

"Well, the sonar's not working and without that, the vehicle is blind. We can't do any useful work down there without the sonar," he said while stating the obvious.

Mark added, "Let me get Ron out here and let's see what we can do with it."

I could not believe it. I thought to myself, "We had just spent four hours diving the vehicle to the bottom and when it gets there, the sonar craps out." Without the sonar, it was a waste of time to try to find any targets. Steve was right. The Magellan was blind and to try and do any searching visually was a waste of time. I sat on an upturned oil bucket and momentarily buried my head in my hands. How could so many things go wrong? Was I resigned to failure on this project? Was Liberty Bell 7 simply destined to remain on the bottom of the ocean? How was I going to explain to the Discovery Channel that we had pissed away over a million dollars with nothing to show for it except for a smattering of unidentified sonar targets? What a waste of time and money!

Mark and Ron dove into the sonar problem with their usual gusto. They were both exhausted, but in spite of that, they quickly began bantering over the communications set as Ron exchanged and tweaked several printed circuit boards housed in the Magellan's nearly empty cable reel.

"How's that, any difference?" Ron queried.

"Nope. . . Still not receiving," Mark answered.

Ron made more adjustments, using an electronics screwdriver to calibrate the potentiometer. "Anything there now?"

"OK, I'm starting to get some returns; give it a few more turns."

As I watched their progress in the Magellan van, the sonar was slowly but surely being resurrected from the dead. After about 30 minutes, it appeared to be in working order.

But I had to judge the situation for myself since I could not see the point in wasting any money driving the Magellan around the bottom without a working sonar.

"Can you let me in there for a minute?" I asked Mark.

I slid into the seat, asking Steve to point the vehicle towards its depressor. I had used this type of sonar for over 20 years and wanted to try this out for myself.

A sharp ringing sound told me that we did indeed have a sonar again. "OK, it's probably not as good as it should be, but it's good enough; start searching." I said.

With that, Mark took over and Steve started driving the Magellan around on a standard search pattern, one where the vehicle was driven out to the full length of its 300 foot long soft tether in an attempt to locate target 71.

As the Magellan ROV scoured the sea floor, the weather continued to build, making it more difficult to keep the Needham Tide on location. The wind was now blasting at about 20 knots, whipping the swells up to a height of around 10 feet. With a ship like the Needham Tide, we did not have long in seas like this. The ship heaved and rolled as the Magellan's steel umbilical swung like a pendulum under the recovery crane. Still the Helmsman fought to keep the ship on her designated spot of ocean.

Steve drove the Magellan out to all points of the compass, finding nothing more than a three foot long section of rusty pipe in the process. He was anxious to get on with it.

"OK, there's nothing in this area, let's move on to the next target." He suggested to Mark.

"Well, let's make sure that there's nothing here, why don't we move the ship 50 meters to the west and search a little more?" Mark answered.

Standing behind them I echoed my agreement, "Yeah, let's check this area out a little more before moving along." That one decision changed history.

With that, the Needham Tide was shifted slightly to the west. The feelings of failure I was having were magnified by what I saw on the monitor. The bottom terrain could not have been worse. We could only see as far as the next sand wave with or without the sonar; maybe 30 or 40 feet at best. I could feel the ship pitching and rolling now, far more than just a few minutes ago. The entire scenario was grim. I could see the future: high seas would force us to recover the Magellan and return to port. And that would be it. The expedition would be over.

"Watch this," said Steve. I looked at the monitor as he plowed the Magellan through the crest of a sand wave, acting as though he didn't have a care in the world. I wanted to yank him out of the pilot's seat and shake him, but I didn't. Soon we would be out of time. *Eighty-eight targets*, I said to myself. *And we won't be able to identify a single one.* I became even more despondent, though I tried to keep it to myself. *Who on Earth did I think I was? I'm not a scientist. . . I don't even have an engineering degree. . . how could I ever think I was qualified to undertake such a venture? Why did I ever think we could find something so small in this terrain?* The thought of having to explain to the news media why we were unable to find Liberty Bell 7 made my stomach do flip flops; like the feeling you get just before vomiting. *Why did Discovery insist on a press conference before we sailed?*

"I have a small return on the sonar," said Mark.

I moved forward. "What have you got?" I asked.

"Hard to say, just some small contacts farther to the west; Steve, head over that way."

The Magellan, bouncing along on a tight tether, probed up a small rise. The targets appeared to be a line of small contacts. *What do we have here?* I wondered.

Into the lights came some light colored debris; a material of some sort. It looked like nothing I had ever seen in 25 years of work on the bottom of the ocean.

"Any ideas?" Mark commented.

"It almost looks like a chunk of ice or something (impossible, of course). Maybe crumpled aluminum from that crashed airplane to the north," I answered.

"Well, there's more of it up the hill and I think I'm getting a larger target behind it all."

"Let's check it out."

The small trail of debris continued as the Magellan clawed its way up a small rise. More pieces of "something."

"What *is* that stuff?" I asked.

"I logged it as aircraft wreckage," Mark replied.

As the vehicle continued to the crest of the hill, I could make out something tall about 30 or 40 feet away. I could not see what it was, just that there was something there. It looked ominous and dark.

As the Magellan struggled westward, the phantom shape materialized before our eyes. *A piece of wing from a crashed airplane?* I thought. I had seen this before on airplane crash sites. But this looked different. . .

"Boy, it's got some height to it," I mumbled. My face became slightly flushed as a wild thought raced through my brain. *You know. . . It's about the right size. . .*

The Magellan's cameras crept closer. The vehicle was bouncing on and off the bottom as it pulled against the tug of the umbilical. My knees weakened slightly. I was almost climbing on top of Mark Wilson's back, as though getting my face closer to the video screen would somehow make the object more visible.

The shape loomed a little closer. My heart rate accelerated until it was pounding with the strength of a locomotive. *Damn. . . damn. . . look at that. . . could that be it? Could it really be?!*

"Oh my god!" I whispered.

I was now making out a conical shaped object with a rough exterior.

Suddenly, the words, "UNITED STATES" burned their way through the glass of the video monitor. Quickly the rest of the capsule came into view, looking like it did that morning in 1961 when it started the 15 minute flight that would put it here, 16,000 feet underwater. It was as though someone hit me with a sledgehammer and thrust the capsule into my face.

My hands covered my face as I exclaimed, "I don't believe it. The first target! This never happens!"

R.D. reminded me, "Sure it does!"

"But it never happens to me!" I yelled, aware of my many failures in life.

"You doubted us!" R.D. responded as he leaped from the control van to spread the word to the rest of the crew. I was shaking with excitement.

All of my problems melted away. I had found what I was looking for.

We had found Liberty Bell 7!

Chapter 8 – Back to the Cape
The Recovery of Liberty Bell 7

There are two great rules in life, the one general and the other particular. The first is that every one can in the end get what he wants if he only tries. This is the general rule. The particular rule is that every individual is more or less of an exception to the general rule.

— Samuel Butler

I stared at the video monitor in total disbelief. It was as though my eyes were receiving the visual input but my brain refused to accept the unreal image of Grissom's capsule sitting on the bottom just as I had predicted for so many years. I was numb with shock. Just 30 minutes after privately writing the expedition off, we had found our target! And it was looking so very good.

Ever so carefully, Steve maneuvered the Magellan closer to our quarry. I could clearly see the open optical periscope flap, the discarded recovery line, the dye marker canister, and landing bag straps protruding from the base of the structure like the petals of a flower. There were a few soft marine organisms attached to the capsule's exterior and some white corrosion on the outside of the recovery compartment, but that was to be expected. As we drove around to the back side of the cold war relic, the words "LIBERTY BELL 7" lit up under our powerful lights as though they were dayglow paint. While Grissom's observation window was intact, we could not see much of the capsule's interior; just some strings of corrosion hanging down from the control panel. The dacron recovery loop was still standing upright and the severed stub of the High Frequency SARAH beacon antenna rested on top of the recovery compartment. I couldn't tell if any part of the beryllium heat sink was there at all, though I thought I glimpsed a few folds of the fabric impact landing skirt. Incredibly, only a meter from the side of the capsule, there was a discarded Coca Cola can (the style of the can indicated that it had not been dumped overboard from our ship).

The Magellan control van became the most popular place on the ship as people piled in and out wanting to catch a glimpse of our historic find. Far below our ship, the Magellan ROV shook, rattled, and rolled, as it cornered the space age artifact, hunting down and documenting every inch of its exterior. Bathed in the Magellan's powerful blue-green lighting, Liberty Bell would have looked like the only object existing in the ink-black void of the ocean deep. The bottom silt erupted as the vehicle's thrusters churned the bottom, maneuvering into position. What sea life there was three miles down, scattered in fear at the sound of the Magellan's high speed electric motor and whirring hydraulic pump. After about 50 minutes of documenting our find, it was time to recover the ROV and "get the hell out of Dodge," as they say. The ship was starting to bounce harder as it heaved in the seas. It was as though I had noticed the weather for the first time. I was rudely yanked out of my 1961 mind set back into 1999, where I was on a ship on the ocean, with the weather building quickly; the sea had given something up. Now it was going to teach us a lesson.

I remember scrambling up the passageway to my cabin to get my notebook and bashing my leg into the door jam in the process. I felt no pain as blood dripped down my leg. Incredibly, after making numerous satellite telephone calls to Discovery and others, the only person I was initially able to reach was Max Ary of the Kansas Cosmosphere and Space Center. "Max, we found it and it looks beautiful. . . you can see the writing on the side and everything," I said.

"Oh boy! Oh boy," was all Max could manage. I thought he was going to start crying on the telephone he was so excited. We all were.

However, we had more immediate problems on board the Needham Tide. Slowly but surely the weather had picked up all day. By the time we found the spacecraft, the winds were gusting to over 30 knots and the seas were at 15 feet or better. This was not a problem if you had a proper dynamically positioned ship. But our Gulf of Mexico "mud" boat was in trouble and having a hard time staying on location. While the ROV's recovery progressed, I reviewed the video tape of Liberty Bell 7 in the sonar van.

Outside, a battle with the elements was raging as the ship's bow was grabbed by the giant fist of the wind and lurched off course. Before we could do anything, the angle of our recovery crane was shoved out of position, causing the triple armored umbilical to leap off the overboarding sheave like a line off a fishing rod. Then the grinding damage started as dozens of feet of the wire strength member were peeled off the exterior of the umbilical like the skin of an apple. Unfortunately, all of this wire had no where to go and bunched up in a huge rats' nest around the traction winch and level wind. By the time the equipment was stopped, the one cable connecting the Magellan to our ship was seriously damaged. As they say in ROV circles we were in deep trouble.

I found out what had happened and started helping by cutting off the loose wire and taping up the umbilical cable to keep it all together. The rest of the Oceaneering crew began the dangerous job of using a grinder to cut off the remaining wire to try and continue the recovery. I knew they would do their best but the main thing was that I did not want to see anyone get hurt. But it was not to be. While this work was going on, the ship took a mighty heave upwards and the Magellan's cable was snapped like so much thread. We lost the vehicle. Now I had to amend my report to Discovery; yes, we found Liberty Bell 7 and by the way, we lost the Magellan ROV in three miles of water. With the Magellan ROV lost, we could accomplish no more at sea. I ordered the ship to head to port, had a well-deserved shot of Jack Daniel's and crashed in my cabin.

When word of Liberty Bell 7's discovery hit the beach, it shot around the world with the speed of Grissom's sub-orbital mission. While I had always thought it would be a big story, the Discovery Channel was somewhat taken aback by the global interest in the 38-year-old space artifact.

Even more surprising was the reaction from the media when they saw the ghostly images of Liberty Bell 7 sitting on the bottom on top of its decomposed heat shield with the white lettering on its side for all to see. As I described our find during a press conference held in Port Canaveral, there was an audible gasp from the reporters as soon as they saw the words "UNITED STATES" beaming from the corrugated side of the Mercury capsule. I can only surmise that their expectations had been tainted by the appearance of the rusting hulk of the Titanic and that they expected to see a barely discernible shape that had once been a Mercury capsule. They were impressed. However, it was what I had figured we would see all along, assuming the craft was intact to begin with.

Unfortunately, we still had big problems with accomplishing our objective as we had a Mercury spacecraft on the bottom along with the Magellan 725 ROV and no way to recover either one. In reality, if it wasn't for the fact that the Magellan was fitted with a state-of-the-art imaging system from Woods Hole, I think Oceaneering would have deemed the Magellan not worth the price of recovery. But we had to get those cameras back.

Who was liable for the loss of the Magellan? Legally, it was Oceaneering, since it was their equipment that was lost while under the control of their operations personnel. However, I certainly felt responsible, wishing I had held out for the ship I knew we really needed. But I was desperate. It wasn't that the Needham Tide could not do the job; it could. But anyone with a significant amount of experience working on the ocean knows that it's best to not cut such margins so close. Due to budget restrictions, we cut it a little too fine and ended up paying for it in the end.

The only other ROV in Oceaneering's stable that could do the job was tied up on a long term contract in the Far East. For a time, I corresponded with Dr. Anatoly Sagalevitch about using one of his Russian Mir submersible to lift the capsule. While the idea of using a Russian submarine to lift a relic of the Cold War was inviting, they didn't have the long kevlar liftline we needed. As a result, we had few other options than to build a new recovery vehicle; one that would raise both Liberty Bell 7 and the sunken Magellan ROV. It would be called the Ocean Discovery ROV, identified as such due to the financial support given to Oceaneering by the Discovery Channel to get the job done. There was no question we were going back; the problem was when.

After all the intense negotiations between Discovery and Oceaneering subsided, what happened next was essentially a crash program to build a 6,000 meter capable ROV to recover the Magellan and Liberty Bell 7. Typically, it takes at least a year to design and assemble such a vehicle. Oceaneering did it in less than three months; an accomplishment demonstrating the expertise of their hard working engineers and technicians in Morgan City, Louisiana. To their credit, Oceaneering also insisted on a Dynamically Positioned (DP) support ship to ensure that we had adequate stability during the operation. To this end, the Motor Vessel Ocean Project was enlisted as a support platform for the capsule's recovery. It was the class of ship we should have had all along.

On a certain level, I still found it almost impossible to believe that we had actually found Liberty Bell 7. After spending almost a third of my life trying to find it, viewing the images of the sunken capsule seemed surreal. Logically, I knew that we had found it, but Liberty Bell 7 had been such an elusive goal for so many years that it seemed impossible that a time would come that I would not be pursuing the "capsule." The numerous dreams I had experienced throughout the years left me unprepared for the actual event. But there it was, on video, the internet, and all over the world for everyone to see. It was as though an enormous balloon had been deflated inside of me. Now that the capsule had been found, what was left for me? What other sea monsters were there left to slay?

I used the time between expeditions to confirm that our recovery tools fit as designed, thanks to the support of the Smithsonian's Paul E. Garber

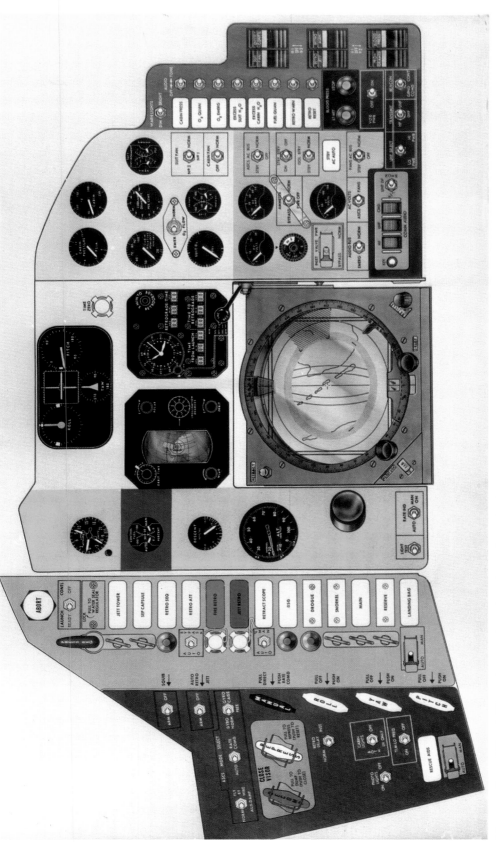

Mercury Control Panel: The main control panel used on Liberty Bell 7 was consistent with the orbital version of Mercury spacecraft, as opposed to the one used on Alan Shepard's Freedom 7 mission. While simple when judged by today's standards, the control areas were color coded and laid out in a logical fashion. The left side of the panel is related to the Stabilization and Control System, the center with the attitude control and navigation displays, and the right side with electrical power and the Environmental Control System. Boeing Illustration.

On the Way Up: Liberty Bell 7 is hoisted to the top of the Redstone booster in 1961. Clearly visible on the spacecraft are the retro rockets (held on by three straps), beryllium heat sink, and capsule window. The spacecraft is being lifted by the same point used during its recovery in 1999, the escape tower mounting ring. NASA Photo.

Night Fire: The massive Ekofisk complex's flare stacks almost turn night into day during inspection operations in 1979. The fact that it's about a mile between the stacks gives some indication of the size of the flames. Working underneath these structures at night turned the deck of support ships into outdoor ovens. C. Newport Photo.

North Sea Recovery: Ocean Systems personnel help TROV S-4 back on deck during inspection and cleaning operations on the Ekofisk Complex's concrete oil tank (right) in 1979. The TROV was eventually fitted with a 9,000 psi jet blaster which was used to clean the surface of the tank for still photography. C. Newport Photo.

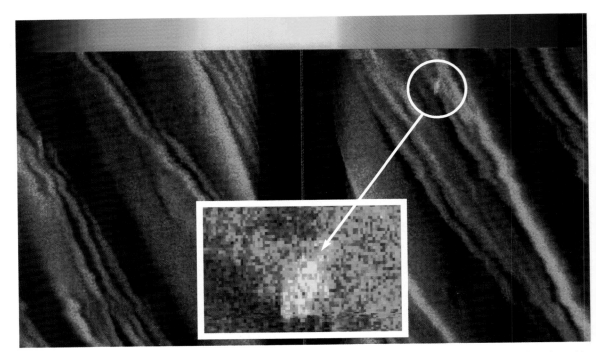

Mysterious Target: It took advances in computer processing software to really advance the capabilities of modern side scan sonars. This image from Triton Technologies QMIPS application shows an unidentified target discovered during the search for Liberty Bell 7. It was correctly identified at the time and later confirmed an the wreck of a wooden sailing ship, lost in 1810. The single blue pixel in the center of the enlarged image at the bottom is a cook stove on the ship's deck. C. Newport Photo.

Air India: The salvage of wreckage from Air India flight 182 off the Irish coast in 1985 was a pivotal undersea recovery operation and did much to convince me that Liberty Bell 7 could be recovered, if it could be found. The SCARAB II ROV goes over the side from the Canadian Coast Guard Ship (CCGS) John Cabot on its way to the bottom in the mile deep waters. Even though I spent over five months straight on the John Cabot, the Air India job remains one of the best operations I ever worked on. Why? Because SCARAB worked. C. Newport Photo.

Sonar Operations: Richard Daley monitor's the Ocean Explorer's surface displays during the search for Liberty Bell 7. The center monitor is the topside "waterfall" display showing the bottom as it passes underneath the sonar. The left monitor indicates assorted diagnostic and operational data. C. Newport Photo

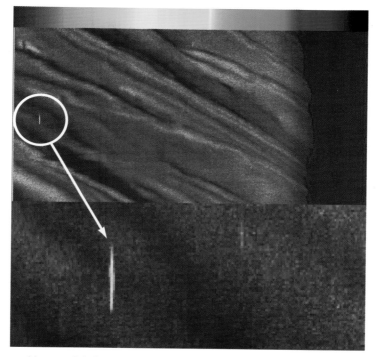

Target 71: There was nothing special about target 71, other than the fact that it was a hard isolated contact right near my best guess of where Liberty Bell 7 actually sank. This particular target was detected on the next to last fill-in search line and was only inside of our search area by about one quarter of a mile from the west. It ended up being the "Holy Grail" of space history. Numerous sand waves are also visible running across the sea floor. C. Newport Photo.

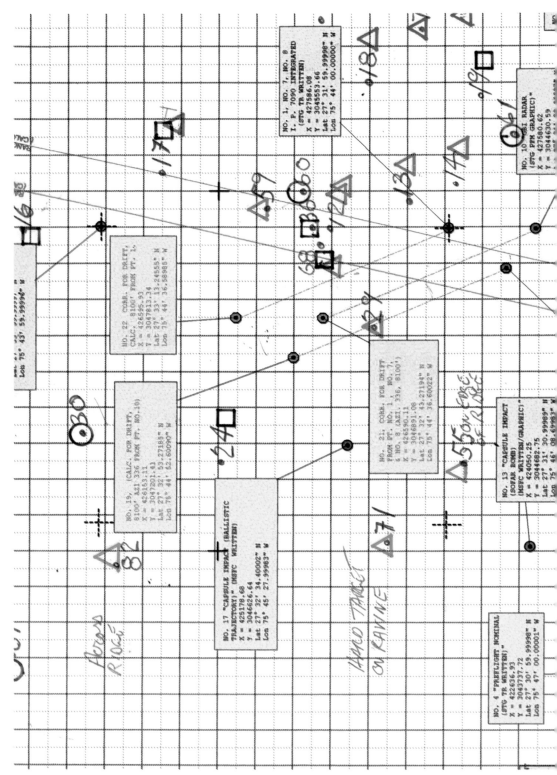

Target Central: This plotting chart shows only a portion of the 88 sonar contacts we had to deal with during the search for Liberty Bell 7. The green circles signify "soft" targets, the squares "medium" hardness objects, and the red triangles "hard" targets. I did this in order to be able to see the big picture and help plan the Magellan's identification dives. The plan for the first dive was to start out with target No. 71, then move southeast across contacts 55, 54, and 28, then east

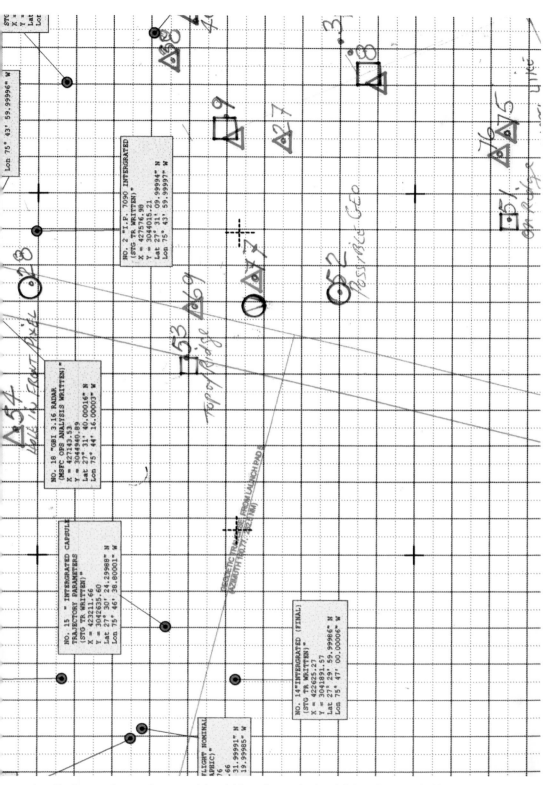

o number 61, then northwest via targets 14 and 13, ending up in the vicinity of target 12. Given the prevailing weather conditions, it's doubtful the Magellan could have stayed in the water long enough to identify all those sonar targets. Fortunately, it wasn't necessary. C. Newport Photo.

On Liberty Bell 7! Lost for almost 38 years, the words "UNITED STATES" glow under our underwater lights as Grissom's missing Mercury spacecraft is finally discovered at 1920 hours, Eastern Daylight Time on May 1st, 1999. Also visible in this view of the capsule is the optical periscope and flap, several landing bag straps, and a huge pile of beryllium oxide material, all that's left of the craft's heat sink. Discovery Channel Image.

Control Van Jubilation: ROV pilot Steve Wright holds the spacecraft in view with Mark Wilson and I looking on, while we reposition the MV Needham Tide. By then, the winds were threatening to push our Gulf of Mexico "Mud Boat" out of position before we had a chance to figure out where we were and where the spacecraft was relative to the rest of the world. Discovery Channel Photo.

Hatch Opening: An even closer view of the spacecraft's open hatch shows the braided stainless steel recovery line, corrosion dangling down from the surface of the main instrument panel, and about a half dozen landing bag straps. The small lever visible just inside the open hatch is part of Liberty Bell 7's Right Console, which controls both suit and cabin temperature. Discovery Channel Image.

Flip Side: Around the backside of Liberty Bell 7 we discovered the capsule's old wire rope recovery line, the famous painted on crack, the capsule's name, and Grissom's open hatch. Part of the vehicle's control panel on the inside is also partly visible. One thing I was worried about was whether or not Liberty Bell 7 even had a backside as it was possible that a capsule-mounted SOFAR bomb had blown a section of the spacecraft away. Fortunately, it was all there. Discovery Channel Image.

Lost At Sea: Not long after this photograph was taken, the Magellan's damaged umbilical snapped in heavy seas sending the valuable craft to the bottom, not far from Liberty Bell 7. At the time, the vehicle was at a depth of about 9,000 feet and these Oceaneering technicians (from left to right, Ron Schmidt, Steve Wright, and Richard Daley) were trying to remove a mass of damaged wire that had jammed up the wheels of the traction winch. Discovery Channel Photo.

Step with Caution: Once the Magellan is in the water, its soft tether has to be hand tended until the vehicle has stretched it out on the surface. Anyone working in the area has to be extra careful to keep from being tripped up on deck or even worse, yanked over the side by the cable. C. Newport Photo.

Garber Mercury Capsule: One of the things I did while the new vehicle was being built was to check the fit of the clamp style recovery tools. Using one of the tools and a test spacecraft supplied by the Paul E. Garber Restoration Facility, I was able to establish where the tools could and could not be placed on the escape tower mounting ring. The red areas signify acceptable tool attachment points and this photograph was used as a guide during Liberty Bell 7's recovery. C. Newport Photo.

In the Water: Oceaneering's Kevin Adams leans hard as he tends the Ocean Discovery's soft tether on the surface. It was Kevin's machine shop in Morgan City, Louisiana that originally built the tools used to lift Liberty Bell 7. C. Newport Photo.

ROV Woes: As with any new vehicle, we had our share of problems with the Ocean Discovery vehicle. Even though the ROV was tested on the way to Florida, most of the problems could not have been predicted and simply go with the territory when using any new hardware. C. Newport Photo.

The End of the Line: After releasing the spooler from the bottom of the ROV, Oceaneering personnel haul it over the ship's stern. The kevlar line on the spooler is a direct connect to the spacecraft. In all, we had about 25,000 feet of recovery line, some of which had to be fed out on the surface so the bitter end of the line could be threaded through the deck mounted traction winch and cable reel. C. Newport Photo.

Keeping Watch: George Brotchi (left) and Kevin Adams (right), help guide the kevlar recovery line onto the ROV's traction winch, which is normally used to pay out and take the vehicle's armored umbilical. From the time we started taking in line, it was about eight hours until Liberty Bell 7 was at the surface. C. Newport Photo.

At the Surface: The recovery vehicle is slowly guided over the stern with the lift line spooler dangling underneath. At this point in the procedure, the end of the line attached to the bottom of the ROV is about three miles away, connected to the top of Liberty Bell 7. C. Newport Photo.

First Glimpse: The upper half of Liberty Bell 7 breaks the surface at 0220 in the morning of July 20, 1999, nearly 38 years to the day after the craft sank. One of my main concerns was that a leg of the nylon sling might get caught under a recovery tool. Fortunately, the capsule came up in good shape. Discovery Channel Photo by Reed Hoffman.

On Deck at Last: Liberty Bell 7's impact landing skirt and straps drag across the ship's fantail as Steve Wright and I guide the spacecraft on deck. It was a moment I never thought would actually happen. Discovery Channel Photo by Reed Hoffman.

Homeward Bound: McDonnell Aircraft's Capsule No. 11 sits on deck just after recovery. Although it's not apparent, the craft's forward hatch had fallen down and was jammed in the side hatch opening. The yellow strap running from the top of Liberty Bell 7 to the right was to help stabilize the artifact for bomb technicians from UXB International. The old helicopter recovery line can also be seen still attached to the top of the capsule. Discovery Channel Photo by Reed Hoffman.

Rat's Nest: When Liberty Bell 7 was hauled off the ocean floor, many components which were barely hanging on to the interior, broke loose and fell into the bottom of the spacecraft. Wiring bundles, periscope parts, and indicator lights litter the water-filled spacecraft interior. The two circular openings in the upper right held the capsule's two low pressure helium spheres for the Reaction Control System. C. Newport Photo.

Space Map: Astronaut Grissom's Mercury Orbital Chart (MOC) and post flight checklist were both found neatly stowed in a fabric flap on the bottom of the optical periscope. The coil cords are lanyards attached to grease pencils used to make annotations on the charts. C. Newport Photo.

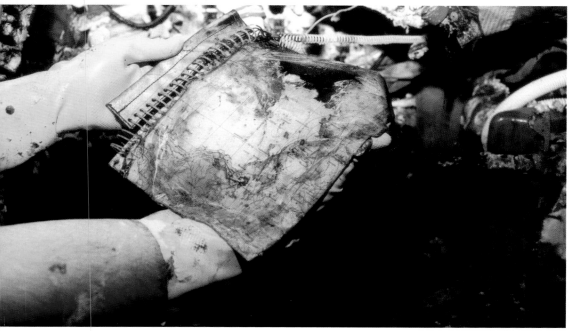

An Unexpected Find: Grissom's Randall survival knife was discovered stuck to the bottom of Liberty Bell 7's aft pressure bulkhead in a congealed mass of muck just underneath the astronaut couch. C. Newport Photo.

Smuggled Money: In all, 52 Mercury Dimes were found inside of Liberty Bell 7, mostly in the seat area and it has been confirmed that they were placed there by the launch crew. It's also theorized that they were originally stored inside of a standard paper roll, probably in the same pouch holding the post flight checklist. When the periscope fractured during recovery, it fell into the seat, spilling the dimes in the process. C. Newport Photo.

Periscope Challenge: Trying to convert items like this optical periscope into usable capsule components is no easy task. Imagery done today using video cameras, computers, and software, were created in 1961 using metal gears and lenses. This item was nearly done in by galvanic corrosion. C. Newport Photo.

Capsule Kit – Some Assembly Required: Removing many of the individual components from Liberty Bell 7's interior was fairly easy. However, it was necessary to remove the capsule's aft pressure bulkhead in order to clean all traces of salt from metal surfaces. To this end, the Cosmosphere's team carefully sawed the spacecraft in two on a line in between the lower edge of the hatch sill and heat sink area. C. Newport Photo.

Titanium Wrinkle: This close up shows a wrinkle in one of the titanium hat stringers which framed the opening for the explosive hatch. There was some initial speculation that this mechanical damage was somehow related to the premature jettisoning of Liberty Bell 7's exit hatch. However, this damage was probably caused when the spacecraft crashed into the water after being released by the recovery helicopter. Film footage taken in 1961 does show Liberty Bell 7 landing in the water on this side. C. Newport Photo.

New Ocean Treasures: Numerous 19th Century artifacts litter the stern of the Atlantic Target (found during the search for Liberty Bell 7), in particular an intact hourglass (center), magnetic compass (left), and several ceramic bowls. Many of these items were recovered and are undergoing cleaning and restoration in Florida. DOE/WHOI Image.

Space Pioneer: Grissom's legacy in space exploration will live forever and even today, his name is spoken with considerable respect. It is not surprising that his grave at Arlington National Cemetery is often decorated with flowers, in particular on the yearly anniversary of the Apollo I fire. I visited Grissom's grave after Liberty Bell was found. I wasn't the only one. C. Newport Photo.

Innerspace Vehicle: I returned to the Blake Basin in the summer of 2000 to identify a mysterious target found during the Liberty Bell 7 search. Two dives to 4,800 meters in this Mir I manned submersible resulted in the discovery of an 1810 merchant ship, code named "Atlantic Target." C. Newport Photo.

Deepest Wooden Wreck: Atlantic Target ended up being the deepest historic wreck ever explored and yielded many fascinating artifacts such as two flintlock pistols, 1,500 silver and gold coins, silk fabric samples, and even a section of newspaper from the year 1809 which was still readable. One of the Russian Mir submersibles is at right hovering just off the bow of the wooden ship. DOE / WHOI Image.

Cockpit: The finished interior of Liberty Bell 7 following restoration. Much of the craft's instrument panel disintegrated during recovery leaving most of the gauges hanging on the ends of their individual wiring bundles. The tinted Plexiglas, which is obviously not part of the original spacecraft, was used to hold each instrument in their original position. KCSC Photo.

Completed Artifact: This side of the completed Liberty Bell 7 spacecraft would normally have been facing towards Earth during flight, as indicated by the protruding lens of the optical periscope. All of the painted lettering on the restored artifact is original and has not been retouched in any way. KCSC Photo.

Astronaut Exit Area: The still intact Vycor glass window, capsule name, and the famous painted on crack are all visible in this view of the restored Mercury spacecraft. The silver material under the window is thermal insulation and only part of the aluminum panels covering the parachute compartment still exist. The circular component just above the hatch opening are the C and S-Band antennas. KCSC Photo.

Restoration Facility. They took the time to remove a Mercury spacecraft from a large wooden crate so I could attach the tools to the escape tower mating ring and document their position with still photography. To show my appreciation for their help, I later copied all of the McDonnell Aircraft Mercury capsule engineering drawings I had for their reference library.

I was on top of the world. No matter how crazy of an idea it had seemed before, we had actually found Grissom's spacecraft. I hungered to get back to sea, lift the spacecraft on deck, and return to Cape Canaveral, with every one of us conquering heroes. I had paid my dues with all the pipelines and oil rigs I'd inspected, telephone cables I'd buried, and mangled aircraft wreckage I'd help drag to the surface. Now it was time to have some fun and actually get some credit for all the hard work; to get some payback.

After a brief press conference in Tampa, Florida, we once again sailed from Port Canaveral for hopefully our last trip to the location. However, this time we had two special passengers on board: Jim Lewis and Guenter Wendt.

Jim Lewis was the same Marine helicopter pilot who made the tough decision to cut Grissom's capsule loose 38 years earlier. Though he left the Marine Corps soon after Grissom's mission, when he sailed with us, he was nearing retirement as the head of Man Systems Integration for NASA's Space Station program. Jim was a soft-spoken individual who was as amazed as any of us that we had actually managed to find Liberty Bell 7 in the first place. I first met Jim in the late 1980's when I was still doing research on Liberty Bell 7. I remembered that during our first telephone conversation, I mistakenly said that he had "dropped," Liberty Bell 7. He quickly reminded me that he had "jettisoned" the spacecraft; a big difference.

Guenter F. Wendt was a legend in his own time as the "Pad Fuhrer" of Projects Mercury, Gemini, and Apollo. He was a lanky guy who spoke with a thick German accent and many said that had he been in charge, astronauts Grissom, White, and Chaffee never would have died in Apollo 1. He was one of the smartest people I had ever met and had seen his share of action by being shot down, not once, but twice in a Junkers 88 aircraft over Berlin during the Second World War (Guenter actually had about 8,500 hours of time as a flight engineer in night fighters). Guenter later penned the book, "The Unbroken Chain," describing his experiences as the pad leader during most of America's manned space program.

Also on board was my friend Max Ary, the director of the Kansas Cosmosphere and Space Center. His museum had supported my efforts from day one and I lobbied hard to have Discovery hire them to restore Liberty Bell 7. The only other option was the Smithsonian, but they probably didn't have the funds to do the work and I doubted any of their people had ever fully disassembled an intact Mercury capsule. The Cosmosphere had.

Finally, two bomb technicians from UXB International, Hugh Sease and Buddy Eanes, were along to remove and dispose of the unexploded SOFAR bomb. No one envied their job at all.

We left Canaveral on the 4th of July and by the 6th were back near what we thought was the capsule site and getting ready to try and find and recover the sunken Magellan 725 ROV. While recovering Liberty Bell 7 was certainly our priority, we decided to go after the Magellan first, for two reasons: First of all, getting the new ROV wet searching for the Magellan would hopefully sort out any technical problems associated with any new submersible vehicle. In addition, if we were able to recover the Magellan first, then we would be able to remove the Woods Hole cameras from that vehicle and retro-fit them onto the new Ocean Discovery ROV. The new vehicle did have a "broadcast quality" camera, but it lacked the high-intensity HMI lights that would give us the best images from the three-mile-deep waters. However, after diving at the suspected location of the Magellan until July 9th, it was nowhere to be found. What was even more disconcerting was the fact that our "new" ROV was experiencing a host of problems ranging from oil leaks to water intrusion into its electrical connectors. I was hoping that these would be sorted out by the time we started looking for Liberty Bell 7.

Once again, high frequency sound penetrated the abyss of the Blake Basin as the gleaming new Ocean Discovery vehicle plodded along the bottom sweeping the area with its scanning sonar. As the technological monster skipped from one sand wave to another, all of us in the control van strained to see something; anything that showed evidence that we had been there before.

Leaning back in my seat, I remarked, "I'm telling you, I don't see any skid marks... or anything else for that matter that shows we're in the right area." Coffee cups, some serving as makeshift ash trays, were scattered amongst the control console.

Ron Schmidt rubbed his weary eyes and continued driving the ROV on its easterly track across the bottom. "Well, we've just got to keep expanding our grid until we find it; it's down here somewhere." Ron grabbed the microphone and called the bridge.

"Bridge. . . ROV. . . Let's move the ship 75 meters at 160 degrees. There's nothing in this area."

"ROV. . . bridge. . . OK. . . it'll take a few minutes. We'll call you when we're there." With that, the navigation and control software on the bridge sent some electrons to the right spots, the ship's thrusters churned, and the Ocean Project slowly slid to the new location. But the capsule wasn't there either. I was getting nervous.

By the next day, our navigators had discovered an error in their calculations and we moved the ship to another spot. Still no spacecraft. How could this be? How could we have been on Liberty Bell 7, taken a satellite navigational fix, and after a day and a half not find it?

By July 10th, we had been at the "capsule location" for almost 30 hours and had yet to find our target. The Ocean Discovery ROV, after starting on another dive, had been recovered due to a ground fault in its rate hydraulic pack. Once again it was on the way to the bottom, a transit that took about four hours each way.

How was I going to explain to Mike Quattrone of the Discovery Channel that we had found the capsule in May and after almost two days on the bottom could not re-locate it?

By July 11th, my patience was at an end. Any confidence that I had previously had in our navigators had evaporated. I didn't think that they had any idea where we should be diving and decided to have my shipmates at Oceaneering figure it out. Fortunately, I had brought along all of the original side-scan sonar records. We had found the capsule before using this information and maybe we could recreate what we had done back in May. Also, Oceaneering had had the smarts to bring along the Ocean Explorer side scan sonar and control van, just in case.

The undulating bottom of the Blake Basin scrolled down the screen on our QMIPS sonar processing computer as I halted the display and positioned the cursor on target 71.

"O.K., there's the target and you can see the eastings and northings we had for the ship when we imaged the capsule." Richard Dailey and Chris Russell, both very experienced operations personnel punched their calculators and scribbled numbers on their note pads.

Chris mumbled a little as he did his figuring, "OK... we were using a lay back of 6800 meters... There's the offset... Let's see here..." More writing and number crunching.

Richard scratched his head, saying, "Jesus... That's over 500 meters from where we've been looking!"

"What!?" I exclaimed.

Chris added, "We're in the wrong spot. We should be about 500 meters to the east."

In the near mountainous terrain of the Blake Basin, 500 meters might as well have been 500 miles.

After moving the ship and in less than an hour, the familiar shape of Grissom's capsule filled our video monitor. It had taken us almost three days; time that I had hoped to use to search for Liberty Bell 7's hatch. Now, that would be impossible. While we were trying to relocate the capsule, I had taken the time to review some of the side scan data and found three targets that I wanted to inspect in the hopes that one of them was the hatch. Even though the tiny hatch was on the threshold of detectability using the Ocean Explorer 6000 sonar (given that we had been searching in the 1,000 meter swath), I knew that the hatch would have landed with the concave side up, intensifying its ability to reflect sound. The targets were also to the east of where we found the capsule, which is where they should have been given the easterly current in the area.

I bounded from the inside of the van stomping my feet on the deck screaming, "Jesus Christ... finally!" Finding Liberty Bell 7 again was almost as good as the first time. When we first discovered the capsule, we did not know it was there and it was an unexpected event. But it had been agonizing to KNOW the capsule was in the area and be unable to get our vehicle to it. We were now back in business.

Both Guenter and Jim were in our control van when we finally got the ROV on the spacecraft; they were dumbfounded.

"Look at that... it's so clean!" Exclaimed Jim, seeing the capsule for the first time in almost forty years. The Ocean Discovery ROV moved up and forward, hovering over the capsule's parachute compartment.

"OK now, that's a good view," added Guenter, "You can see the parachute liner we'll have to pull out to get at the SOFAR bomb."

We continued to document the area where we planned to connect our recovery tools by flying the ROV around the small end of the capsule; we already had one tool on the ROV so we would be ready for work.

I had established the general recovery procedures years ago and was now going to see first-hand how they worked. On this, our inspection dive on the capsule, we planned to get a better look at the escape tower mating ring and attach one of the recovery tools (each one clamped around the inside and exterior of the ring like small vices). We would then attach at least two more tools, then carry down to the bottom a recovery line spooler holding about 25,000 feet of 3/8" kevlar line. Once the spooler was on the bottom near the capsule, the free end of the line would be attached to a three-point lifting sling (the sling connects all of the tools together). Finally, the spooler would be hauled to the surface underneath the ROV and placed on deck. The remaining kevlar line would be fed out on the surface until the end was routed through the ROV crane and traction winch (the part of our deck gear that fed the ROV umbilical in and out) and stored on the vehicle's cable reel on top of the ROV's armored umbilical. The last task would be to haul in the kevlar line until the capsule was near the surface and ultimately lift the whole thing on deck. Our original plan also incorporated a large nylon cargo net which we hoped to deploy on the bottom and use as a "safety net" while lifting the capsule to the surface. However, after talking it over with Oceaneering's George Brotchi, we decided that working with the net on the bottom would be nothing but trouble. I decided to put my faith in our recovery tools and McDonnell Aircraft's strong design. Another thing that concerned me was the water that would be contained within the capsule as it cleared the surface. The added weight of the water forced Jim Lewis to jettison the capsule in 1961. Would the added weight cause the spacecraft to fall apart after it was lifted clear of the surface? There was no way to know.

One problem was that the custom-made recovery tools could not just be placed anywhere on the capsule's escape tower mating ring due to the numerous fasteners and sensors sticking up around the perimeter of the ring. Consequently, the photographs I had taken at the Garber facility were our guide to exactly where each tool would be installed on the capsule. The biggest unknown was the strength of the connection between the ring and the rest of the spacecraft. I thought it was pretty good considering that during a typical abort scenario, the capsule would experience about 20 gravities of acceleration, which meant that that mounting ring had to be firmly attached to the rest of the spacecraft. But how strong would those rivets be after being submerged in salt water for almost four decades? Even so, I was convinced that raising the capsule by that ring offered the best opportunity to get it back intact with as little damage as possible.

I also wanted to attach a 37 kHz pinger (an underwater acoustical beacon) to the capsule before we did the lift, in case we lost the capsule during recovery. Unfortunately, the type of sonar fitted to our new ROV was incapable of detecting these devices.

The Ocean Discovery ROV hovered near the top of the encrusted parachute compartment, right manipulator outstretched, moving in for the kill, so to speak. Clamped in the right claw was the first of three recovery tools we needed to connect to Liberty Bell 7 so it could be hauled to the surface and from its 1961 grave.

I thumbed through the Garber images, pointing at one showing a tool attached to a sister capsule.

"OK, Steve, you're looking good, try to hook the edge of the tool on the ring in this area," I said.

Steve Wright lit up another smoke and nudged the vehicle ahead. About three miles below us, the robotic contraption slid closer to our objective. If we could have heard any sound, there would have been a slight clunk as he snagged the inside of the escape tower ring with one edge of the tool's teeth. The capsule was now hooked by one edge of the recovery tool.

"You're looking good Steve, all you need to do now is rotate the jaw and start tightening it up," I whispered.

"Well, it's hard to keep it stable. . ."

"You're doing good, man, just keep turning. . ."

Far below us, the ROV's right claw struggled to twist the tool onto Grissom's spacecraft as the screw turned even tighter.

Finally, after a few minutes, the recovery tool informed us that it had had enough as its rotation ceased. It was tight. Only two more to go.

After some nagging problems with our new ROV, we managed another dive on July 13th and connected two more recovery tools to the capsule as well as the three-point lifting sling; all done in less than five hours. With that, we backed the vehicle away from the sunken capsule and plopped down on the bottom to grab Liberty Bell 7's dye marker canister, which was resting on the rusting end of a small wire rope amongst the landing bag straps. The aluminum fingers foraged for the canister, grasped it, and plucked it from the base of the capsule. Fortunately, the connection between it and Liberty Bell 7 had long since dissolved due to the corrosive action of the sea water. Liberty Bell 7 was ready to be lifted.

Both myself and Max Ary stood on the rolling deck of the Ocean Project as the Ocean Discovery ROV swung under the recovery crane. I leaned under the vehicle and grabbed the canister.

"OK, let it go!"

A technician inside the control van pressed a red button. The jaws opened as the canister fell into my hands. The first piece of Liberty Bell 7 had been recovered from the deep after 38 long years.

Max and I examined our find and after a few photographs, dunked the still shiny canister into a cooler full of fresh water. Green dye burped from the cylindrical object.

"Look at that," I exclaimed, "the thing's still putting out dye!"

Max, as surprised as I was, added, "Well I'll be damned."

We were on a roll. However, our "roll" was stopped dead in its tracks by a series of technical problems spanning almost a full week.

While I liked the way the spacecraft was rigged, I was concerned about what would happen to the whole mess once we started lifting. I didn't see any problem as long as the lifting sling stretched out straight and pulled on the individual tool lifting rings. But I knew from past experience that there was no way the spacecraft was going to be lifted gently from the bottom. As soon as all the slack in our lift line was taken up, the spacecraft, along with all of our attached rigging would be jerked violently upward in a series of bounces. In addition, there was a good chance that Liberty Bell 7 would be dragged across the bottom for some distance before it actually cleared the sea floor. But my biggest fear was that one leg of our three point sling would become hooked underneath one of the tools (the edges of each tool stuck out from around the side of the capsule) ripping it off in the process and weakening the whole structure. Once we started pulling, we were doing it blind because the ROV had to be on deck before we started lifting.

Our biggest problem was that our new ROV was breaking down before my eyes. Everything seemed to be going wrong now. The CCD camera was out, the low light SIT camera was on its last legs, the vehicle's telemetry system was intermittent, and we had problems with the ROV's power supplies, compass, electrical connectors, as well as numerous electrical shorts to ground. Then the surface sonar display crapped out. That killed us because now we needed parts; parts we did not have. Consequently, in short order we were steaming towards Marsh Harbor, near Great Abaco Island to pick up these critical components.

By this time I was definitely feeling the strain of the operation, not that there was any other way. I was not getting enough sleep, my rest being limited to cat naps at all hours of the day and night. The pressure of the job was forcing me to smoke way too much and I had also developed a rasping cough. I sounded horrible. When George Brotchi, Oceaneering's Project Manager, started worrying about me, I knew I was in trouble. I was handling it all as well as could be expected but it was extremely frustrating. Here we were with the capsule all rigged to be lifted and all we had to do was carry one kevlar line to the capsule, snap it onto the sling, and haul in. But we couldn't do it because our recovery vehicle was broken, and broken good. The only thing that had gone right was that the weather had been workable the whole time. But that could change in an instant.

Marsh Harbor was a welcome respite from the fatiguing tedium of shipboard life. The ship's captain managed to negotiate a good price with a local water taxi and it was not before long that we were all headed into the beach to get good and drunk. Every one of us deserved it. I admit to not remembering for sure, but I think Guenter stayed on board. But Jim Lewis, ever the Marine, carried the NASA flag well and had a blast. Semper Fi. We virtually took over the only decent bar that was open and bored the help with sea stories of dubious truth. Some bought Tee shirts for their wives, daughters, or whatever. However, most of us wandered the darkened streets of the small village looking for something, anything. But even with that, we always ended up at the one bar whose name I will never remember. We were just happy to be off the ship and away from the pain.

After leaving our mark on Marsh Harbor, we sailed on July 18th loaded to the gills with sonar parts and a slightly hung-over crew. It would be my fifth trip to the Liberty Bell 7 site and I figured that by now I could find the capsule by smell without a satellite navigation system.

It was about that time that Peter Schnall, the director of the documentary film for the Discovery Channel, approached me about exactly when the capsule would be raised; in other words, would it hit the surface in daylight or at night?

Peter explained, "This is the money shot. . . the million dollar image. . . we need to film it during the day because the lighting will be so much better."

I knew what he was saying but after all we had been through, I could not see delaying the recovery one minute. Plus, I had my orders.

"Look," I replied as I pulled out a piece of paper, "This is a fax from Mike Quattrone, the head of the Discovery Channel telling me to raise the capsule as soon as possible. You get him to send me a fax telling me to wait for daylight, I'll do it. Otherwise, he's writing the checks on this operation and until I'm told otherwise, as soon as we're ready, the capsule is coming up."

As we steamed back towards the capsule, it was interesting to observe both Jim Lewis and Guenter Wendt as they adapted to ship board life. Though I had met both of them before, I really didn't know either one of them all that well; two individuals who had very different but pivotal roles in Grissom's sub-orbital mission.

Jim was a little quiet at times, but always cheerful and friendly to anyone he spoke with. He would spend the hours reading books, watching movies and seemed to be enjoying his vacation from his office at the Johnson Space Center; he was ready to retire. Given what I saw of his piloting skills during the aborted recovery of Liberty Bell 7, he probably would have made an excellent ROV pilot.

Guenter was all business and spent much of his time trying to calculate the breaking strain of our kevlar recovery line and compile timelines on our operation; he was trying to figure out when we would be done. He was a real nuts and bolts individual and would have been a good operations manager. One thing was certain, he knew more stories about the space program and astronauts than anyone I had ever met. Because of all the delays, Guenter also missed the 30th anniversary of the Apollo Moon landing because he was stuck at sea on our ship; I felt horrible about it. All of his buddies were in Florida living it up while he was at sea. However, I hope they understood and at least he would get the opportunity to view firsthand the recovery of Grissom's Mercury capsule.

It wasn't until July 19th that the Ocean Discovery ROV was able to reach Liberty Bell 7 again after one aborted dive; we were all hoping that our technical problems were behind us. Hampered by the recovery line spooler dangling under the vehicle, Ocean Discovery bounced across the bottom doing an excellent job of dropping the spooler near the sunken capsule; too good in fact. The spooler was so close to the capsule that we were concerned that the spacecraft might be hit by the ROV as it rigged the line to the capsule. So with great difficulty, the heavy spooler was lifted clear of the bottom and deposited about 50 feet away from the space age artifact.

Needless to say, by this point in time I was physically exhausted, felt like death warmed over, and anxious to get the whole thing over with. While everything that was happening provided more moments to be savored, the finding and recovery of Liberty Bell 7 had dragged on for so long that I looked forward to the moment that I could put it behind me.

The vehicle's manipulator mounted hydraulic motor whined three miles below us in the darkness as the final connection to Liberty Bell 7 was made. A thick steel pin was screwed into a circular opening linking Grissom's capsule to the Ocean Project via several miles of space age kevlar line. Far overhead, the ship's propellers chewed at the warm Atlantic Ocean waters keeping our recovery platform in place. I found it ironic that much of the technology we were using was a direct result of NASA's space program. If our country had not committed itself to going to the Moon, would we have had kevlar, miniature computers, broadcast quality color video cameras, satellite navigation systems, and the ability to form titanium into impregnable electronics bottles? Maybe not.

Finally, after a few minutes it was done. Liberty Bell 7 was now poised to be yanked from the year 1961 into 1999, whether it liked it or not. Its 38 year long flight was going to finally end. There was an undeniable connection between the day that Grissom fought for his life in the ocean and our struggles to return Liberty Bell 7 to the American people.

As the Ocean Discovery ROV was hauled to the surface it deployed the kevlar line and I wondered what would we find when the capsule was on deck. Would it hang together as I had predicted? Or would the escape tower ring be ripped from the top of the spacecraft leaving us scratching our heads for how to get the thing on deck.

Three hours after the final dive by the Ocean Discovery ROV, after feeding out 4,000 feet of line down current on the surface, after rigging the free end of the line in the deck mounted traction winch and cable reel, and after starting up the diesel engines powering the heavy deck gear, we started taking in the kevlar recovery line. We did it slowly at first, in order to soften the initial strain on the capsule. But I knew that in reality, the raising of Liberty Bell 7 would be anything but gentle.

By 2100 hours we figured that all of the slack in the line was gone and we had to be taking a strain on the capsule. I visualized what was happening on the bottom, but tried not to think about it.

Liberty Bell 7 felt the first stirrings of recovery in the form of gentle tugs on the recovery line as it danced above the spacecraft like a long white snake. As we continued to haul in on the line, the top end of the capsule would start moving, first with a few small jerks, them becoming more pronounced as more slack was taken out of the line. Finally, there would be a mighty lurch as the full weight of the capsule was borne by the line, exploding Liberty Bell 7 from it's nearly four decades of rest in a cloud of powdery beryllium oxide and swirling bottom sediment. As more and more line was taken in on the surface, the capsule would be dragged across the crests of nearby sand waves, swinging like a pendulum and bouncing up and down under our ship, until it reached a state of equilibrium in the deep ocean currents. That was what I thought was happening. The reality of it, though, was unknown to me. And somewhere far below us, part of the legacy of Virgil I. Grissom dangled dangerously on the end of a line about the size of my little finger.

Forty-nine minutes before the 30th anniversary of the Apollo 11 Moon landing and one day before the 38th anniversary of the launch of Mercury Redstone 4 with Liberty Bell 7, the diesel engine running the powerful hydraulic winches squealed, coughed and ground to a halt. Now, as we scrambled to figure out the problem, Liberty Bell 7 started bouncing dangerously on the end of a very long string.

"What's the problem?" I shouted in the darkness of the night as Steve Wright and a few other Oceaneering crew peered into the entrails of the engine.

"Looks like the alternator shaft bearing has gone. . . we'll have to replace the whole thing." The shaft bearing had eaten itself to bits grinding up the shaft in the process. An hour later and following furious work by the crew, black smoke once again spewed from the engine's exhaust pipe as the winches started slowly rotating again.

The dripping kevlar line continued to thread itself onto our deck gear like a massive ball of string until we interrupted the recovery again as it appeared to go slack. Guenter and Jim watched nervously from the upper deck.

"They've stopped!" Exclaimed Guenter. "I wonder what's wrong?"

Jim stared silently at the scene before him.

But it was only a trick of the eye and sea. While the recovery line looked like it no longer held any weight, in reality, it did. The illusion was caused by a combination of the dynamics of the capsule on the end of such a long line and the timing of the heave of the ship. Every so often the timing of the two against each other made it look as though the line was slack; but the firm press of a steel-toed boot against the line confirmed that the capsule, or whatever was left of it was still there. At least for now.

By 0200 hours, I was on the fantail of the Ocean Project pacing like a nervous parent with the wail of the deck gear in the background. I strapped on a life jacket, some gloves, and a hard hat, and readied myself for the inevitable.

Peter Schnall was poised with his expensive 16 mm film camera and the deck was ablaze with lights. At that moment, it seemed as though we were the only pinpoint of life in the blackness of the Atlantic Ocean. A hoard of people made up of Oceaneering and Discovery personnel and the ship's crew gathered at the stern in anticipation of the recovery. At 0210, we transferred the load from the smaller 3/8" line to a larger one inch line and started the final lift. While the smaller line was strong enough to get the capsule to the surface, we were concerned about the weight of the capsule in the air, along with any water contained within the titanium pressure shell of the spacecraft. Whatever was still down there was less than 100 feet from the surface.

This was it. For over 14 years I had fantasized about this moment. The moment when a dripping wet conical object would be hauled out of the sea; when it would be exposed to the atmosphere for the first time since the year 1961; when there would no longer be one Mercury capsule missing at sea; when they would all be on dry land; and when Liberty Bell 7 would be returned to Florida, to the United States, and back into the world.

My first glimpse of the object of my obsession showed itself at 10 minutes after two in the morning, looking like an apparition from times past. The dark night gave the object a surreal appearance, the cylindrical parachute compartment bursting through the surface amidst chunks of white corrosion toppling off the sides of the small end of the capsule. From all appearances, it looked like we still had a firm grip on the spacecraft, though one of the recovery tools hung on at a disconcerting angle; it obviously had been loosened during the long trip to the surface.

I should have been thinking about the 14 years of hard work it had taken to get to this point, but I wasn't. All that was going through my mind was, ". . . Let's not drop it now." Ever so slowly, the Magellan's recovery crane swung the capsule, only the top of it exposed, across the stern to the clear area where Steve and I stood ready to take possession of my dream come true. Thoughts of concern and anticipation shot through my brain with the speed of Grissom's 1961 space mission.

The capsule steadied itself as the line slowly raised it from the sea.

"C'mon baby. . . Just a little longer. . ."

The capsule continued it's upward journey as rivers of sea water cascaded down the sides of the cone shaped vehicle.

"Easy now. . ."

The words UNITED STATES glowed under the force of our spotlights.

"Bring it in now. . . Just a little more. . ."

I lunged over the stern of the ship getting my first touch of the capsule. My hand slipped off of the bottom part of Liberty Bell 7. I laid my body on the deck to see if any part of the heat shield was still attached. It was not.

"Damn. . ."

Liberty Bell 7 rotated slightly as both Steve and I gripped the edges of the hatch opening that Grissom had scrambled through 38 years earlier.

And ever so gently, Liberty Bell 7 ended its 38 year long space voyage as it splashed down on our wooden deck at 20 minutes after two in the morning of July 20, 1999.

"We got it!"

In a moment of resignation to the finality of our accomplishment, my body hunched over until I heard behind me the cheers and clapping of the entire ship. I rotated around with excitement and started clapping in acknowledgment of their enthusiasm and thrust both arms into the air in a moment of triumph. It was finally done.

With McDonnell Aircraft's capsule number 11 safely onboard we quickly anchored it to the deck with numerous nylon straps, mostly to immobilize it for the bomb technicians from UXB International. They would need as much stability as possible before starting to search for the unexploded SOFAR bomb.

But even though the capsule was still a dangerous artifact, people gathered around it to examine our find.

Guenter quickly made his way to the stern, placing his outstretched hand on the still wet corrugated shingles saying, "The last time I touched it was 38 years ago. . ."

Jim Lewis was in awe of Liberty Bell 7's condition, remarking, ". . . and it's a good looking spacecraft." It almost looked as though Guenter was laying his hands on the dripping capsule to purge it of its sins; hell, whatever it takes. . .

While Liberty Bell 7's condition was indeed remarkable, the interior was a mess. The forward hatch had come loose from its mount and was lodged in the explosive hatch opening. I gingerly extracted it from the capsule and finally had my first look at the interior. Water continued to drip all through the inside as I stuck my head in, seeing that part of the once intact control panel had disintegrated, leaving numerous flight instruments dangling like apples on a tree.

But what struck me the most was the smell... It was like the odor of carbon or decayed wood, probably from the chemical action of the electrolyte in the craft's batteries. The optical periscope had broken in half and was now laying amongst other rubble such as the decomposed remains of the control panel and what was left of the astronaut camera. I could see part of one of the film spools lying exposed which eliminated any hope of saving the film.

Hugh Sease and Buddy Eanes moved in quickly to secure the area so they could get to work. I tried to help.

"I removed the forward hatch so you could see better inside... you want to have a look?" I asked Hugh.

"We're not doing anything until you get all of these people out of here."

With that, I waved my hands to get everyone's attention saying, "We need everyone out of here NOW... Everyone's got to go right now!"

People took a few more photos and slowly started to move away below decks so if the bomb exploded while Hugh and Buddy were removing it, the crew would be protected by the steel structure of the ship; if the worst happened, however, Hugh and Buddy would be killed or at least severely injured. According to UXB, the safety exclusion area for the SOFAR bomb was 900 feet and we were on a ship little more than 200 feet long.

As Hugh started rummaging through the muck at the bottom of the capsule's interior, I went to relative safety on the bridge to answer a telephone call from Discovery (even I was prohibited from staying on deck).

Gesturing at the still dripping capsule, Hugh commented, "Well, it's either got to be up there (pointing at the parachute compartment) or down there (indicating the area near the open hatch); I'm going to look down there." Hugh poked his head into the smelly interior and started foraging for the explosive device. As he did this, Buddy continued to spray the exterior of the spacecraft with fresh water in an effort to keep everything wet. Their efforts were recorded by a remote video camera mounted above the spacecraft.

Dripping wet fingers dug at the liquefied remains in the bottom of Liberty Bell 7; a mass of tangled wires here and there, most of them attached to assorted switches, gauges, and circuit breakers. Hugh did it mostly by feel, shoving aside the shattered astronaut observer camera and decomposed remains of the control panel. But in a few minutes, his hand gripped a hard cylindrical object about the size of a beer can, which was coated with deep sea muck. He pulled it out of the capsule to examine it in the spotlights beaming down across the deck. Buddy splashed some water on the object. As the coating of mud washed away the word, "BERMITE" was revealed.

"This is it!" Hugh remarked.

As I continued with my telephone call on the bridge I received the word and hastily ended the call to get back to the fantail.

Hugh held his prize for me to see saying, "This is armed but unexploded."

I snapped off a few photographs with my Nikon and after examining the device a little more told them to toss it over the side. After the expedition, there would be some people who thought we should have tried to disarm the device as it was part of the history of the capsule. But SOFAR bombs were not designed to be disarmed in the first place and the best place for it was at the bottom of the ocean. It was another headache we could do without. But what got me were the odds of it all. The SOFAR bomb was specifically designed to mark the location of a sunken Mercury capsule. And we had just recovered the only such spacecraft that had indeed sunk and the bomb was a dud.

Now that Liberty Bell 7 had been rendered safe, Max Ary and I began our documentation and post recovery work in earnest. While both of us took a few photographs of the capsule's interior, it was obvious that there was little we could do with it until the spacecraft was back at the Cosmosphere. I grabbed a bilge pump and started dewatering the bottom of the capsule as Max poked around inside to see what was really there. In short order, Max pulled his hand outside, gave me a nudge, showing me several shiny objects in his palm.

"Well I'll be, Max exclaimed with excitement. "Mercury dimes."

"But I thought all of the dimes were inside of Grissom's space suit?"

"These must have fallen out or been placed there by someone else; I'll also wager that we're going to find a lot more of them in there too."

Max handed the three dimes to me which I stuffed in my pocket as we continued to work (the dimes were later returned to the museum). It was then that I noticed potential trouble.

"Max, look here, some of the lettering is starting to peel off."

Sure enough, part of the number "7" and the "L" from Liberty Bell 7 were started to peel up from the one of the corrugated shingles like paint that had been left in the hot sun for many years.

I added, "We've got to get this thing underwater and out of the air before we lose the rest of the lettering."

With that, I got Steve and the rest of the Oceaneering crew to carefully lift the capsule off the deck as we slid a large nylon cargo net underneath. Getting the spacecraft transferred from the stern to the Cosmosphere's transport container proved to be a harrowing experience, as the spacecraft, encapsulated within the cargo net, swung out of control over the ocean as the Ocean Project rolled from side to side (the tag lines we had attached to the capsule proved nearly useless). The only way to control the capsule's motion was to dunk it back into the ocean to dampen out the swinging as I literally put my body, twice, between the capsule and the side of the ship. But somehow we got the spacecraft into the bottom half of the heavy steel container, which was manufactured by the Superior Boiler Works and designed specifically for transporting one Mercury capsule.

After seeing the problems we had with simply getting Liberty Bell 7 into the container, I was having serious reservations about lifting the heavy container cover and placing it over the capsule at sea.

As the Ocean Project slowly steamed into the seas, I expressed my concerns to Max.

"Look, that container weighs several thousand pounds and I would not try to do this at sea; there's a real good chance that it will crash into the spacecraft doing serious damage."

Max acknowledged my concerns saying, "We've got to get the capsule out of the air. . . If we leave it the way it is it may suffer irreparable damage until we get back to Florida."

We had come so far in recovering the capsule and I was hesitant about risking damaging it at this point in time. But using numerous tag lines and simple brute force, we managed to get the top on the container without hitting the capsule. It was not long before we had gallons of water filling the container and expunging the corrosive salt air from it.

By then I was both physically and mentally exhausted. I had been up for over 35 hours and could not concentrate on anything. The best I could do to walk was to shuffle my feet about the ship until I finally collapsed in my cabin, which I shared with Max.

Later on in the morning, as the Ocean Project steamed towards Port Canaveral, I gathered the disgruntled Oceaneering crew (ROV crews are always disgruntled) on deck and expressed my appreciation for all their hard work. I also encouraged all of them to sign the outside of the capsule's transport container documenting their involvement in the mission, the deepest commercial salvage operation in history. I popped the cork on a magnum bottle of Moet champagne and we celebrated as much as our one bottle allowed.

In darkness of the morning of July 21, 1999, a single beam of light from a television news van illuminated our ship as we returned Liberty Bell 7 to a location not far from where the tiny capsule had been launched 38 years to the day. Some reporters commented that we had planned it that way. But anyone who was on board the Ocean Project during the recovery would have known that it was not so. With all of the twists and turns of fate during the expedition, we could not have planned our next meal, never mind when we would recover Liberty Bell 7. We were happy just to have done it at all. I thought about all the years I had spent trying to recover Liberty Bell 7 and considered if it was worth it. I suppose it was. But after all of it, the most satisfying event was after the capsule had been off-loaded from our ship and chained down to a flat bed truck destined for the state of Kansas, and I was asked by the driver to fill out a bill of lading. I still have a copy of it now and part of it reads: "Transport container holding McDonnell Aircraft Capsule Number 11, known as Liberty Bell 7."

Chapter 9 – Epilogue

Nothing recedes like success.
— Bryan Forbes

A lot of things happened following Liberty Bell 7's recovery. After the spacecraft arrived at the Cosmosphere in Hutchinson, Kansas, it sat inside of the transport container and was flushed with fresh water for several weeks. Max Ary, the President of the museum, said that during one month their water bill was a staggering $9,000. In the meantime, the museum's restoration team, headed up by Vietnam veteran Greg "Buck" Buckingham, worked long hours to finish up their restoration facility, basically a walled-in area on the bottom floor of the museum fitted with an oversized shower stall big enough to hold the spacecraft. The area was also outfitted with numerous work benches and a window for visitors to watch the work as it progressed.

There's no question that simply having the recovered spacecraft at the Cosmosphere was a tremendous windfall for the little known museum. In fact, Ary later commented that their visitorship increased by over 40% and the museum's webcam, aimed at Liberty Bell 7, received an incredible 550,000 internet hits in December of 1999 alone (the Cosmosphere's internet web site was visited over 2 million times that year). Certainly, public interest in the missing spacecraft ran high throughout the United States, as well as the rest of the world. In fact, truck driver Max Davis, who drove the recovered spacecraft from Florida to Kansas, later said, ". . . as I hauled the Liberty Bell across the United States, people honked, waved, and took pictures of the huge cargo tank." They even used modified sign language indicating that they knew what was inside the steel container, by acting like they were ringing a bell, then holding up five, then two fingers; the result, Liberty Bell 7.

I flew to Hutchinson to help the museum staff remove the spacecraft from its container and assist with the initial activities. Jim Lewis, the Hunt Club 1 helicopter pilot in 1961, was there as well and I don't think he would have missed it for the world. He didn't waste any time getting up to his elbows in capsule muck. We got the top of the container off without any problems and attached two of the remaining recovery tools to the capsule's escape tower mounting ring, lifting the capsule out with a crane, just as it had been raised from the bottom of the ocean (of the four recovery tools, two remain with the Cosmosphere, one is on tour with the spacecraft, and one was left on the bottom). Even though they had done nothing but run fresh water through the tank for several weeks, there did appear to be noticeable deterioration in comparison to what I remembered seeing when we sealed the artifact up on the dock at Port Canaveral. While this was surprising, there was nothing we could have done about it and we managed to get the capsule transferred onto a forklift and lift it up into place in the museum's large fresh water shower facility so we could begin investigating the interior and removing numerous small items from the inside.

Taking everything into account, Liberty Bell 7 was in remarkable condition. The spacecraft's basic titanium load bearing structure looked like it could have flown again; it was still shiny and except for the fact that all of the air-filled panelettes (part of the double wall cabin structure) had been squashed, it was a breathtaking sight. About 60% of the aluminum panels covering the parachute compartment had corroded away due to galvanic action with the capsule's titanium structure. The flat black colored nickel-steel alloy shingles on the spacecraft exterior were nearly perfect and it did not look like we had lost much more paint on the craft's lettering and famous crack. The impact landing skirt, except for a couple of small rips, was intact. The interior, however, was another story. Everything that had corroded heavily ended up in the bottom of the spacecraft creating about a two foot slurry of water and bits of dissolved control panel. The control panel, from the periscope to the left side, along with assorted circuit breakers, switches, etc., was intact. However, the right side of the control panel from the periscope on simply didn't exist anymore. All that was left were numerous gauges, toggle switches, and dials, dangling on the ends of their wiring bundles like the leaves of a tree. Of the two low pressure helium gas spheres (for the Stabilization and Control System), only one was still there and it had been popped by the tremendous water pressure as the capsule sank. However, both of the grapefruit sized titanium high pressure oxygen spheres were intact. The optical periscope was broken in half and had fallen into the astronaut's seat. I had not expected that much corrosion on the aluminum components, but there it was.

Even with the above, there was much to impress everyone involved as we explored the guts of Grissom's capsule. The foam padding on his seat was totally intact, complete with darkened outlines where his restraint harness had rested for almost four decades. Rummaging around in the capsule's "bilge," I discovered Grissom's Randall survival knife and the cap to the explosive hatch actuator, with still readable lettering saying, "EXPLOSIVE HATCH IGNITER." The knife was only slightly rusted, attached to a small coiled lanyard; I found it stuck to the aft pressure bulkhead by a solidified mass of gunk. The emergency flashlight, after getting a new bulb and batteries, worked like new. Grissom's orbital maps and post flight checklist were both readable and we could even see the grease pencil markings the astronaut made on the checklist after splashdown. His Mylar life raft still held air. There was still fresh water inside of his can of emergency drinking water. I was able to remove the stainless steel fasteners holding the exterior shingles to the titanium cabin with a Phillips head screwdriver and a small hammer. All it took was a few taps and they came out. I do not believe that a single one of those fasteners had to be drilled out.

While in many ways, my job was over, I had very distinct ideas about what the Cosmosphere should do to Liberty Bell 7. First of all, we all agreed that the fact that the spacecraft sank and spent many years on the bottom of the ocean was a integral part of its heritage. Consequently, what the Cosmosphere planned to do was "preserve," as opposed to "restore" Liberty Bell 7. All of the physical attributes that were unique to this spacecraft were to be left as is. While everything was going to get cleaned, as much as possible, nothing was going to be repaired or made like new. However, there were practical decisions that had to be made. For example, it didn't make sense to display the spacecraft with most of its gauges hanging on the end of their wires. I suggested that they view the spacecraft as the bones of a dinosaur: put everything back where it was supposed to be and develop a way to hold various components in their original location. Hence, the missing areas of the capsule's control panel were recreated using material that was obviously not part of the original spacecraft. That way, people could see where things were suppose to be and it would be obvious what was original and what was not. The paint and lettering would be cleaned and preserved, but left as is. I also felt very strongly that the capsule should be displayed on its side with the landing bag deployed, because that's how it was lost (and recovered). Max Ary, being an agreeable sort, always listened to my suggestions and for the most part, went along with them.

We also found a lot more Mercury dimes scattered around the interior of the spacecraft, most of them discovered in the couch area. I later theorized that they had originally been held inside of a standard paper coin roll, which had long since decomposed. The coin roll was probably stored in the fabric pouch attached to the bottom of the optical periscope. When we lifted the capsule from the bottom the periscope broke in two, falling into the seat and spilling the dimes. In total, 52 silver dimes were found (by several people in fact). Cosmosphere technician Dale Capps made another amazing find when they further disassembled the spacecraft: Carefully wrapped inside of a small plastic tube (hidden within a wiring bundle) were five silver-certificate dollar bills, one of them signed by Gus Grissom. One of the others had signatures from some of the launch team and a crack drawn across George Washington's portrait, along with the words "Cap #11 Launch Crew." There were four names, which while hard to decipher, look like "V. Monrow, Geo. Baldwin, Darrell Boren, and C. Powers." Fortunately, no one's yet asked for their money back. Jim Lewis found a cigarette butt and the remains of a standard plastic drinking cup (it was in a layer of sludge below and to the left of the control panel). Where they came from and how they got inside a supposedly "clean" spacecraft remains a mystery, though Guenter Wendt remembers having a small water cooler and drinking cups in the white room before launch.

Buck Buckingham of the Cosmosphere had a challenging task: he had to figure out how to disassemble and clean a complex space vehicle which was not designed to be taken apart, and do it without degrading the artifact's special history. Once we had finished emptying out most of the putrid fluid filling the capsule's bottom, we carefully removed the survival kit on the left side of the astronaut couch and emptied the contents into several water filled containers. Many of the individual items in this area were coated with a fluorescent green dye, obviously a secondary dye marker for the astronaut's personal use. A lot was also made of the bar of Dial soap found in the survival kit. Why would Grissom need such a thing for a 15 minute flight? I figured that it was part of the standard US Air Force survival kit used on Mercury missions. Later on, I told Max Ary that it was included because, "space flight is a dirty business."

All during this period, the spacecraft was kept under a light shower of fresh water to keep everything moist. This biggest problem was in identifying the hundreds of "parts of things" that had ended up in the bottom of the spacecraft. Obviously, a component database was going to be critical to keeping track of every nut and bolt removed from Liberty Bell 7. Max Ary initially estimated that the spacecraft disassembly would involve working with 35,000 separate components, but later revised this number to about 25,000 due to the number of things that simply weren't there anymore. Many minute parts were found by straining granular material through a screen into buckets and in some cases, what looked like a simple chunk of sodium, when carefully broken open, would have a switch or small fluid valve contained inside, like the jewel of a crystalline geode. Local dentists donated old dental tools and instruments to help clean thousands of tiny mechanical components.

Cleaning every square inch of Liberty Bell 7's physical structure was absolutely critical to making sure that no further corrosive damage would be done to the priceless artifact once it was put back together again. Many unpaid local Hutchinson volunteers (some experienced with restoring classic automobiles and aircraft) toiled for months doing this very detailed and painstaking work. However, getting the capsule apart was not easy. The initial plan was to remove the spacecraft's aft pressure bulkhead to gain access to the interior, since working in the cramped confines of the cockpit was almost impossible (McDonnell Aircraft originally assembled these spacecraft from the inside before this bulkhead was installed). Unfortunately, after removing several hundred nuts and bolts from this area, it was discovered that it was spot welded in place and not designed to be removed after installation (the restoration team destroyed hundreds of titanium drill bits in the process without success). One way or the other, the capsule's interior had to be opened up so it could be cleaned using sand blasting or plastic beads until every speck of corrosion was removed. Eventually, the Cosmosphere reluctantly decided to cut the capsule in half, in between the lower edge of the hatch sill and the heat shield area. This was done using a power saw and at last the interior sub systems were able to be removed from the capsule's basic structure. Fortunately, it was a clean and even slice.

Surprisingly, the Cosmosphere almost forgot about the seven channel tape recorder (manufactured by Consolidated Electrodynamics Corporation) which was mounted to the interior of the spacecraft and they were able to locate and remove the item before the reels of magnetic tape dried out. Sadly, the 4,800 feet of Mylar based tape still sits in a small tub of water as the Cosmosphere has yet to find a sponsor to cover the cost of restoration, estimated at about $40,000. Somewhere on that tape may be words (also possibly the sounds of a hatch exploding from the side of Liberty Bell 7) spoken by a 1961 Gus Grissom that were not transmitted to the ground during his flight, if only someone would pay to get them back. The film from the two D.B. Milliken film cameras was destroyed over the years when those two items either imploded from excess water pressure or corroded away.

All of the Cosmosphere's work was done under the watchful eyes of museum visitors and anyone with a computer and an internet account could log onto the "Liberty Bell 7 Cam" at any time. I used to check out the Cosmosphere's progress on a daily basis to both keep tabs on their progress and remind myself that the capsule's recovery was real, and not imagined. During all of my time working on the project never in my wildest dreams did I ever think I would be able to turn on my computer and see Liberty Bell 7 on dry land.

Using an assortment of dental picks and more than 800 dremel tool brushes, the Cosmosphere restoration team very carefully removed every bit of corrosion from the capsule components. They even took apart the circuit breakers and toggle switches, digging out the white salty material from the metal parts. Once these electrical components had been cleaned, lubricated, and reassembled, they still worked. It was extremely difficult to figure out where some of the smaller parts, no longer part of anything, even came from. The spacecraft's hand wound satellite clock was reduced to a pile of tiny brass gears, cleaned, and reassembled. The seven miles of electrical wiring and connectors were made to look almost like new and tied back together as during the original assembly of the space vehicle in 1960 in St. Louis. The numerous electrical and pressure gauges, some having glass faces shattered by water pressure, were taken to bits and put back together, some with new faces. Originally, I did not think the gauges were hermetically sealed – apparently, some of them were and when the capsule sank, the glass faces shattered. Fortunately, the broken glass was still in place. The capsule's triple plane vycor glass observation window, the first such component ever flown in space, was cleaned and ended up looking like new. Only one of the three panes was damaged. One question I still have is what happened to the paint surrounding the observation window? It had to have been painted (or anodized?) and this area, while originally having a flat black color, is now bare metal. However, the left two letters of the words "United States," the "U" and the "S," are still there, even though they were painted on after that section of the capsule was coated. If what's under the letters is gone, why not the letters themselves? Did the paint somehow protect the coating in that area?

Grissom's last ditch survival tool, his nylon personal parachute, after being hung up at a local fire station, was cleaned with fresh water and repacked in its original container by a parachute rigger in Hutchinson. Sadly, the post flight checklist filled out by Grissom after splashdown, did not survive the cleaning process and the writing was consumed by the paper's acid content.

However, photographs of it remain showing the astronaut's original grease pencil markings. One problem the Cosmosphere had was trying to get the salt water out of the thousands of minute cells contained within the aluminum honeycomb structure of the astronaut couch. Each time they drilled into the item, salt water squirted out, still under significant pressure. Eventually, though, the corrosive liquid was extracted from the seat before it was reinstalled into the spacecraft. Slowly but surely over several months, Liberty Bell 7 was reassembled as the Kansas volunteers carefully put everything back into their original locations. Once all of the individual components were back in place, the titanium aft pressure bulkhead was reattached to the conical afterbody structure using stainless steel plates and fasteners. When the exterior shingles were added, the fact that the capsule had been cut in two was not even apparent to anyone looking at the artifact. The restoration workers even reset all of the capsule's toggle switches back into their original 1961 positions. After the work was completed, Liberty Bell 7, even with most of the underwater induced damaged left as it was found, was a remarkable artifact, destined for a three-year, 17,247-mile Discovery Channel sponsored tour of the United States' finest museums. By the time the Cosmosphere was done, it had taken 6,500 man hours and cost an estimated $250,000 to bring Liberty Bell 7 back to life.

Of the seven original Mercury astronauts, only four of them lived to see Liberty Bell 7 recovered. The spacecraft's loss didn't have a negative effect on Virgil I. Grissom's career at NASA and in spite of what some perceived as "unanswered questions" about why the capsule sank, Grissom went on to command the Gemini III mission thereby becoming the first man to make a second trip into space. Along with astronaut John Young, Grissom, the "pilot's pilot," changed his spacecraft's orbit, even performing some complex banking maneuvers during reentry to guide his space vehicle towards the carrier USS Intrepid in June of 1965. There was no question that NASA had faith in Grissom's abilities, and for this reason gave him command of the Apollo I spacecraft. Unfortunately, his hopes with the Apollo program and a crack at a Moon landing were not to be; Grissom, along with astronauts Ed White and Roger Chaffee, lost his life on January 27, 1967, during an oxygen-fed electrical fire in their Block I Command Module.

It was ironic with Grissom. If the one thing that caused him to lose Liberty Bell 7 had been installed on his Apollo spacecraft, he would probably be alive today. What I find unbelievable about the Apollo I fire was that in 1967,

NASA concluded that filling the interior of a complex spacecraft (with all of its electrical wiring, switches, circuit breakers, batteries, etc.) with pure oxygen was an acceptable risk. At the time, one fact was certainly common knowledge in commercial diving circles: high partial pressures of pure oxygen represent a serious fire hazard in any hyperbaric environment. Even before the mid-1960's, no competent diving company would have ever dreamed of filling a deck decompression chamber up with 100% oxygen and putting three men inside. *Never.* Yet, our premier scientific agency in the United States misjudged a technical issue known to any qualified diving supervisor in the Gulf of Mexico. Apollo I, as well as the later Space Shuttle Challenger disaster, have hopefully taught NASA once and for all that basic physical laws apply to everyone.

Deke Slayton, Grissom's best friend during the Mercury, Gemini, and Apollo programs, died in June of 1993. After NASA yanked him from flight status during Project Mercury, he finally got his ride into space during the Apollo-Soyuz Test Project (with astronauts Tom Stafford and Vance Brand) in 1975. In his book, "Deke!" he made an important point about Grissom when he wrote, "One thing that would probably have been different if Gus had lived: the first guy to walk on the Moon would have been Gus Grissom, not Neil Armstrong." Alan Shepard, after sitting out projects Mercury and Gemini due to an inner ear problem, was the only Mercury astronaut to go to the Moon and commanded the Apollo 14 mission. Shepard departed the Earth on his final mission in July of 1998.

Mercury astronauts Glenn, Schirra, and Cooper all flew in space again after their Mercury missions. John Glenn, after a long hiatus from space travel, as a United States Senator from Ohio managed to convince NASA to give the 77 year old former astronaut a job as a Payload Specialist on STS-95, where the agency used his historical medical data to conduct "space-based research on aging." It's not for me to question whether or not it was worth it, but the publicity surrounding Glenn's return to space gave NASA a much-needed shot in the arm. Maybe Glenn deserved it after getting cheated out of any more flights by President Kennedy, who didn't want to risk a national hero's life on another trip into space. Cooper flew with Pete Conrad on Gemini V in 1965, a long duration eight-day flight which further pushed the design limits of the Gemini spacecraft. Wally Schirra, commanded the Gemini VI mission in 1965 supported by Tom Stafford and after a nearly disastrous start (their Titan II booster ignited then immediately shut down during launch), accomplished the first true space rendezvous using Gemini VII (piloted by Frank Borman and Jim Lovell) as a target vehicle. Schirra also led the first successful Apollo mission in 1968 accompanied by Walt Cunningham and Don Eisele proving that the redesigned complex Command Module was safe to fly. These days, the surviving Mercury astronauts travel across the country supporting promotional activities or working with the Mercury Seven Foundation in Florida.

Sadly, two individuals key to the success of the overall project are no longer living: Greg "Buck" Buckingham, the Cosmosphere's Chief Restoration Craftsman and George Brotchi, Oceaneering International's Project Manager on the second Liberty Bell 7 expedition. They were both good men.

When the Liberty Bell 7 exhibit made what was termed a "dazzling debut" at the Kennedy Space Center Visitor Center on June 18, 2000, I was halfway around the world in the Philippine Sea searching for the Heavy Cruiser USS Indianapolis on another Discovery Channel sponsored expedition. Unfortunately, unlike the Liberty Bell 7 operation, numerous equipment breakdowns made it impossible to complete an investigation of my search area and the famous ship remains undiscovered. When talking about the Indianapolis, what I tell people is that if we had only searched half the planned area looking for Liberty Bell 7, we wouldn't have found that either. Maybe I used up all my good luck on Liberty Bell 7. I'd still like to find the Indy, if only I could find someone to pay for it. Mercury astronaut Scott Carpenter, himself experienced in underwater operations, had kind words for the Liberty Bell 7 project at the exhibit's opening saying, "Space flight is brief and glorious, deep ocean exploration is dirty and cold and really hard. I have a great respect for those who found and recovered Liberty Bell."

Me? First of all, what most people don't know is that Liberty Bell 7 was not the only thing found in the Blake Basin in 1999. Just outside of our search area to the north, we also imaged what looked like an old sailing ship. In the year 2001 and with the support of Mike McDowell and his company Deep Ocean Expeditions, I returned to the location where Liberty Bell 7 sank and made two dives down to 4,800 meters in the Russian manned submersible Mir I, operating from the Research Vessel Akademik Keldysh. While it wasn't the valuable Spanish wreck we had hoped for, we did find an amazingly well preserved 1810 merchant ship. Overall, we recovered silver and gold coins (about 1,500 in all), as well as numerous artifacts including a wooden sextant and octant, two flintlock pistols, an intact hourglass, as well as a silk fabric sample with the words "Not to be Sold," still visible. Incredibly, we also found a still readable newspaper from 1809, wrapped around 13 gold coins inside of a small snuff box. Had the wreck turned out to be a bust, we had planned to spend our time searching for Liberty Bell 7's explosive hatch. Unfortunately, the famous door remains undiscovered to this day. It can probably be found (with great difficulty) but it would cost a fortune and maybe it's not even worth it. After all, what's the point? We got the capsule back and isn't that enough?

Surprisingly, even though my dives in Mir I were not to its maximum depth, more humans have traveled into space than have been to the depth from which Liberty Bell 7 was recovered, about 4,800 meters, or roughly three miles down. In addition, following the record breaking dive in Trieste I in 1960 by Don Walsh and Jacques Piccard, not a single person has revisited the Challenger deep. Not one. Why? Because of the cost involved. Dr. Robert Ballard says that any more dives in the area would only be a publicity stunt and that, ". . . there's nothing down there but mud." I disagree. How can we possibly know what's down there after a total of only 20 minutes at the deepest known part of the ocean? We cannot. If someone had only made one trip to the New World, would we consider it explored?

To put things into a financial perspective, these days, depending upon who does the accounting, one Shuttle flight costs about $400 million. Given the typical charter cost of a 6,000 meter capable side-scan sonar or ROV, the money used to do *one* Shuttle launch, could pay for *36 years of deep water operations*. In that time, you could survey an area over three miles deep that was *nine times* the size of the state of Texas, or an area about half the size of the European continent. Who knows what we might find down there. An organism holding the key to a cure for cancer? For Aids, perhaps? And you could get all of that for leaving the Shuttle on the ground only once.

Don't get me wrong. I'm not suggesting that we scrap the Shuttle program in favor of ocean exploration, or even scale it back. I'm only saying that for the most part, less is known about the deep ocean than the environment of space (or the Moon for that matter), for the simple reason that less money is spent on undersea exploration. Our annual budget offers an interesting insight into what our country's priorities are: NASA's 2002 budget is $14.5 billion, which is over four times the funding for the National Oceanic and Atmospheric Administration, which will receive only $3.3 billion for the same year. Of course, if you really want to see what our tax dollars pay for, the Department of

Health and Human Services gets to play with a staggering $459.4 billion for the year 2002 (that's almost half a trillion dollars). Incredibly, some people still think that *more* money should go to social programs. You could build a whole fleet of underwater vehicles for what they probably misplace in accounting alone. Consequently, it doesn't look like the United States is all that interested in exploring space *or* the deep ocean. It makes you think.

When James Earl Jones said during the Liberty Bell 7 documentary film that I was putting my career on the line, at the time I thought it was just dramatic verbiage. Now I've discovered that there was a lot of truth to that statement and in some ways, I'm paying for it now. Why? Because when you get right down to it, I'm just an ROV pilot, granted, an experienced one, who managed to find and recover a 1961 Mercury spacecraft. And with the ongoing economic downturn, times are tough even for "underwater explorers." However, I'm happy to have fulfilled what was really just a dream and when you look at the big picture, it's a miracle we were able to pull it off at all.

A lot of people, including some of my peers, have suggested that I just got lucky. The problem is that there's both good and bad luck. I didn't feel very lucky when we flooded the sonar's electronics bottle not once, but twice, using up three valuable days of search time. Losing the Magellan ROV didn't feel lucky either. I also didn't feel all that lucky when it took us over two days to find the spacecraft again on the second expedition. However, some things did work out in our favor.

Based on what I now know of the subsurface currents in the area where the capsule sank, it's possible that Liberty Bell 7's actual splashdown point in 1961 was slightly west of my search area (the spacecraft was within our operations area by less than half a nautical mile). But as the capsule sank, the strong easterly currents took it into the area of the ocean floor we examined. The sand waves we plowed over with Mir I in 2001 were even taller than I first imagined and some of them rose as high as almost 50 feet above the sea floor. If Liberty Bell 7 had been masked by one of those features, we never would have found it. Not that it could not have been found, but taking into account the time we had on site, there's no way we could have searched the area in such detail because we simply didn't have the money. What about the SOFAR bomb? This device had one purpose and one purpose only: to mark the location of a Mercury spacecraft that had sunk to a nonrecoverable depth. And out of all of the SOFAR bombs manufactured, the one fitted to the only Mercury capsule that actually sank was a dud. What were the chances of that happening?

The positioning of our search area had nothing to do with luck. It was established on the basis of my past research, technical analysis, and instinct. In addition, we dove the Magellan ROV on a particular group of targets because they looked good, from the acoustic standpoint, and were closest to my best estimate of where Liberty Bell 7 sank. In other words, my gut instinct told me that was where we should start. What was it about target number 71 that made me pick it as the first one to dive on? There was nothing in particular, except that it was a good hard, isolated return and it made sense from the operational standpoint to look at it first. Out of a total of 88 sonar contacts, there were 11 that we thought were promising. These were the ones deserving of two check marks on our target identification sheets and before the capsule's discovery, Liberty Bell 7 was described as a "hard target across ravine." What about Liberty Bell 7 being the first target? I have never had any problem with good luck.

Maybe it was time for Liberty Bell 7 to come home. Maybe there was some unknown force that guided our sonar in the freezing cold of the deep ocean. Maybe the spirit of Gus Grissom, wherever he is now, gave us a hand. I have no idea. The main thing is that I did what I set out to do and no one will ever be able to take that away from me.

— Author's Note —

I wrote this book for two purposes: First, to create an accurate account of how Liberty Bell 7 was found and recovered and second, to illustrate to others that ideas can be turned into reality if a person is willing to make the necessary sacrifices. And believe me, when you eat, sleep, and dream a singular goal, you make a lot of sacrifices. It took a long time to get this account published, partly due to my involvement in another underwater expedition the year following Liberty Bell 7's recovery. In addition, it was not so easy to find anyone willing to publish the story to begin with and for that I an indebted to Rob Godwin and his company, C.G. Publishing, for believing that I had an important story to tell.

What I tried to do in this work is focus on Grissom, Liberty Bell 7, and the search and recovery expeditions. As a result, while I do discuss Grissom's background and Project Mercury, from the timeline standpoint, I don't generally discuss events after the Mercury Redstone 4 mission. There are a few exceptions, but I considered most topics on space flight activities after July 21, 1961 outside of the scope of this book. Overall, I tried to incorporate historical information on space flight and project development with respect to how it all related to Liberty Bell 7. There are many other books already published that do a far better job than I could ever attempt in describing the early history of American's manned space program and I see no point in duplicating those efforts. One thing I did do though was go into some technical detail on how a Mercury spacecraft was built and operated, mostly because I felt it was important, and not something previously written about in detail.

There are many people who made this undertaking a reality, some who directly participated in the two expeditions and others who did not. First of all, if Mike Quattrone at the Discovery Channel had not supplied the financing to bankroll the capsule's finding and recovery, Gus Grissom's Mercury spacecraft would most likely still be on the bottom of the ocean. Mike believed in me and my idea and he certainly shares in the credit for what we eventually accomplished. When I first met him and others at Discovery, I was just some guy with an idea; granted, it was one that I had worked on for many years, but it was still just an idea. He took a hell of a risk and I'm glad he had the guts to do that. While I was always as honest as possible about our chances of success, ultimately, it's impossible to predict the probability of accomplishing anything on the ocean. I was also very lucky to have others at Discovery willing to work with me on the numerous complexities that make up any deep water salvage operation; in particular, Sarah Hume, the project's production manager, Karen Baratz, who helped the media understand what we were doing, and Tom Caliandro and Michael Prettyman, both integral to getting the project off the ground at Discovery.

One thing that was good about working with the Discovery Channel is that, for the most part, they never tried to second guess me. In other words, they did their job and left me alone to do mine. Granted, like anything else, there were realistic limitations on what we could afford to do. However, unlike other comparable expeditions, I didn't have some wealthy financial backer standing behind me who fancied himself a lay undersea expert. For better or worse, I was left to my own devices, and from the perspective of a successful undertaking, it was the smartest thing Discovery could do.

Also, Max Ary at the Kansas Cosmosphere and Space Center was an early project supporter who was able to think "outside of the box," as opposed to the numerous other historical institutions in the United States who simply did not have the imagination to see what was possible. The Cosmosphere supported me when others didn't want to be bothered with crazy ideas about recovering 1961 space capsules. The museum's Greg Buckingham also accomplished a near impossible task in taking Liberty Bell 7 apart and getting it back together again.

In addition, Maryland-based space history expert Gregg Linebaugh introduced me to a lot of astronauts when I first started working on the Liberty Bell 7 project such as Thomas P. Stafford and Gene Cernan, both of whom did what they could to support the idea. They didn't have to do anything, but unlike most people, they helped as much as their busy schedules would allow. Jim Hartz, the former NBC space flight commentator and Today Show host, also gave me much valuable insight into how to structure and sell the project.

There were also many other retired and active NASA and contractor personnel who tried to fill in the technical blanks about Liberty Bell 7's loss and possible condition such as Dr. Max Faget (who also took the time to review my manuscript), Robert F. Thompson and Jerry Hammack of the Space Task Group, John Yardley of McDonnell Aircraft, Jim Lewis, the pilot of the primary recovery helicopter (and later of the Space Task Group and Johnson Space Center), Robert H. Gillock and Captain H.E. Cook, both formerly of the USS Randolph, Sam Beddingfield, Guenter Wendt, and especially Ray Silvestri of the Johnson Space

Center, who spent much time reviewing my navigational studies to make sure I was not totally nuts to think Liberty Bell 7 could be found. Also, Glen Swanson at the Johnson Space Center History Office assisted me during the writing process. I would also like to thank Tom Hanks and Sooki Raphael at the Playtone Company for supplying the book's introduction. I know there were many others out there and to anyone who assisted my research efforts, I sincerely appreciate the help. It's hard to find people these days who are willing (or qualified) to discuss the accuracy of FPS-16 tracking radars, SOFAR bombs, and suborbital integrated trajectories, if you know what I mean.

Many of my former associates at Oceaneering International's Advanced Technologies Division put in incredibly long hours supporting the expedition such as Mark Wilson, Ron Schmidt, George Brotchi, Steve Wright, and other field personnel, all members of the Magellan 725 and Ocean Explorer 6000 operations team. Also, Chris Klentzman, Oceaneering's former manager of government operations in Upper Marlboro, Maryland, helped by letting me take the time off to pursue the project and Godik Gyldenedge in Oceaneering's Houston, Texas office, showed enormous patience with everyone as the project geared up for operations.

Also, Peter Schnall, Jerry Risius and Glen Marullo, the film crew from Partisan Pictures, were consummate professionals during both of our harrowing sea voyages. I know they've done it all before, but I appreciate them helping an amateur like me not make too big a fool of myself on camera.

There were also numerous divisions of NASA that helped me collect all of the detailed historical and technical data, such as the history office in Washington, as well as those at the Kennedy Space Center, George C. Marshall Spaceflight Center, and Johnson Space Center. NASA's Office of the General Counsel in Washington was a tremendous help in sorting out all of the legal issues related to the capsule's ownership – nothing would have happened without their valuable assistance.

Finally, I sincerely appreciate the public support given to the Liberty Bell 7 project after the fact by Lowell and Norman Grissom and Wilma Beavers. It was a pleasure to get to know their famous brother through their personal recollections of his career.

I know there are others out there who helped over the years and to all of them, I say thank you.

Finally, any opinions expressed in this book, unless otherwise noted, are mine and mine alone. I have taken great pains to recreate the dialogue spoken while we were at sea, but there was not anyone writing it down and if I got anything wrong, I apologize. I have simply tried to tell it as I remembered it. Also, while I have made an effort to make all technical aspects of the book as accurate as possible, I do not consider myself a world expert on Project Mercury history or the design of a Mercury spacecraft. With respect to working on and in the ocean, nobody knows it all, though some know more than others. If others disagree with anything I've said within these pages, I guess they'll have to write their own book to tell their side of the story.

Curt Newport
Potomac, Maryland
July 2002

— Appendix A —
Project Mercury Manned Flight Data

Mercury Redstone Flight No. 3 (MR-3)	
Pilot	Alan B. Shepard
Spacecraft	No. 7 – *Freedom-7*
Launch Vehicle	Redstone No. 7, delivered March 30, 1961
Date Launched	May 5, 1961
Maximum Altitude	116.5 statute miles
Maximum Speed	5,134 mph
Orbits	N/A (Sub-Orbital)
Flight Duration	0:15:28
Remarks	Primary Objective was to evaluate man in space; first United States manned space mission

Mercury Redstone Flight No. 4 (MR-4)	
Pilot	Virgil I. "Gus" Grissom
Spacecraft	No. 11 – *Liberty Bell-7*
Launch Vehicle	Redstone No. 9, delivered June 12, 1961
Date Launched	July 21, 1961
Maximum Altitude	118.3 statute miles
Maximum Speed	5,168 mph
Orbits	N/A (Sub-Orbital)
Flight Duration	0:15:37
Remarks	Second United States manned space flight with a similar objective as MR-3. Spacecraft was lost at sea during recovery. Recovered July 20, 1999.

Mercury Atlas Flight No. 6 (MA-6)	
Pilot	John H. Glenn
Spacecraft	No. 13 – *Friendship-7*
Launch Vehicle	Atlas No. 109-D, delivered November 30, 1961
Date Launched	February 20, 1962
Maximum Altitude	162.2 statute miles
Maximum Speed	17,554 mph
Orbits	3
Flight Duration	4:55:23
Remarks	First United States orbital flight to evaluate man in orbit. Flight was shortened due to concerns over a loose heat shield indication (later proved to be false).

Project Mercury Manned Flight Data
(continued)

Mercury Atlas Flight No. 7 (MA-7)	
Pilot	Scott Carpenter
Spacecraft	No. 18 – *Aurora-7*
Launch Vehicle	Atlas No. 107-D, delivered March 6, 1962
Date Launched	May 24, 1962
Maximum Altitude	166.8 statute miles
Maximum Speed	17,549 mph
Orbits	3
Flight Duration	4:56:05
Remarks:	Second United States orbital flight with objectives similar to MA-6. Astronaut Carpenter overshot landing point during reentry into Earth's atmosphere. Carpenter was later located and retrieved by Navy search aircraft homing in on the capsule's radio beacon.

Mercury Atlas Flight No. 8 (MA-8)	
Pilot	Walter Schirra
Spacecraft	No. 15 – *Sigma-7*
Launch Vehicle	Atlas No. 113-D, delivered August 8, 1962
Date Launched	October 3, 1962
Maximum Altitude	175.8 statute miles
Maximum Speed	17,558 mph
Orbits	6
Flight Duration	9:13:11
Remarks	Objective of the flight was to evaluate man and machine in orbit for nine hours.

Mercury Atlas Flight No. 9 (MA-9)	
Pilot	Gordon Cooper
Spacecraft	No. 20 – *Faith-7*
Launch Vehicle	Atlas No. 130-D, delivered March 18, 1963
Date Launched	May 15, 1963
Maximum Altitude	165.9 statute miles
Maximum Speed	17,547 mph
Orbits	22.5
Flight Duration	34:19:49
Remarks	Flight object was a manned 1-day mission in orbit. The MA-9 mission was the longest of Project Mercury.

— Appendix B —
Specifications, Manned Instrument Satellite Capsule
(MAC No. 11 / MR-4)

General	
NASA Designation	Project Mercury
Designer's Name	McDonnell Aircraft Corporation (MAC)
Model Designation	Model 133K
Number and Places for Crew	One cabin enclosure
Launch Vehicle	Redstone Missile
Mission	Launching of the capsule into a ballistic trajectory having a range of approximately 191.6 nautical miles, an altitude at apogee of approximately 112 nautical miles, and total flight time of approximately 15.8 minutes.

Dimensions	
Launch Configuration	Length – 311.46 inches (with escape tower)
	Diameter – 74.5 inches
Orbital Configuration	Length – about 134 inches (without escape tower)
Landing Configuration	Length – about 91 inches (without deployed landing bag)
	Length – about 139 inches (with deployed landing bag)

Weight	
Launch Configuration	3,983.75 lbs.
Coast Phase Configuration	2,798.68 lbs.
Floatation Configuration	2,272.59 lbs.

Weight and Balance Summary	
Structure	605.41
Adapter – Capsule to Booster	117.64
Escape System	1061.19
Heat Sink	342.57
Stabilization Control System	266.97
Retrograde System	286.79
Landing System	282.79
Instruments and Navigation Equipment	106.97
Electrical Group	258.07
Communications	109.28
Environmental Control System	122.02
Telemetry and Recording	87.24
Recovery Gear	30.48
Astronaut and Provisions For	249.17
Ballast	57.16
Gross Weight Launch Vehicle (lbs.)	3983.75

Specifications, Manned Instrument Satellite Capsule

Pyrotechnics	
Escape Rocket	Solid propellant rocket (Grand Central Model No. 477-80100) motor with three nozzles canted nineteen (19) degrees from the longitudinal axis of the rocket case. The nominal action time for the escape rocket was 1.39 seconds with an average resultant thrust of 52,000 pounds at its center line.
Posigrade Rockets	Three (3) Model D20763 Atlantic Research solid propellant rockets having a total vacuum impulse of 475 pound-seconds providing an average thrust of 370 pounds each for an action time of 1.35 seconds providing a separation velocity of 32 feet per second
Retro Rockets	Three (3) Model TE-316 Thiokol solid propellant rockets with a velocity decrement of 500 feet per second, a total vacuum impulse of approximately 13,000 pound-seconds providing an average thrust of 992 pounds each for 13.2 seconds
SOFAR Bombs	Two Radioplane Model No. 101010 SOFAR (Sound Fixing and Ranging) bombs each containing 11 ounces of HBX high explosive, one ejected at main parachute deployment (set for 2,500 foot depth) with the other permanently mounted to the capsule structure and used to transmit sound ranging signals at a depth of 3,000 feet underwater.

Recovery Aids	
Parachutes	Two nylon 63-foot diameter reefed (12% for 4 seconds) parachutes (main and reserve) providing a stabilized sinking speed of 30 feet per second at 5,000 foot altitude for a 2,160 pound capsule.
Impact Skirt	Fiberglass / silicon rubber-impregnated impact skirt attached to the capsule forebody area and heat sink at 80 points equally spaced around the heat sink.
Fluorescent Dye Marker	Metal perforated dye marker canister ejected with the reserve parachute after landing.
Recovery Flashing Light	High-intensity flashing recovery light which illuminated at a rate of 15 flashes per minute and visible for 50 nautical miles on a starlit Monocles night.
Rescue Beacon	An HF / UHF MCW / pulse modulated unit containing 243 megacycle SARAH rescue beacon and 8.364 MCW portion of a SEASAVE beacon. The HF beacon has a power output of 1.0 watt and the UHF beacon total power is 7.5 watts.
HF Rescue Voice Communications	Amplitude-modulated HF transmitter-receiver using the same basis modules as the HF orbital voice communication system with an output power of 1.0 watt

General Construction	
General	Semi-monoque construction consisting of a conical and cylindrical sections using titanium and titanium-vanadium alloys for load bearing structure. Conical section consists of an unbeaded inner skin seam welded to a beaded outer skin with 24 equally spaced longitudinal stringers. Two bulkheads form the pressurized cabin area. Cylindrical section has a single skin with 12 equally spaced stringers and internal shear webs supporting the parachutes.
Heat shield	Dish-shaped heat sink fabricated from hot pressed QMV grade beryllium, forged to a final size of 74.5 inches diameter with a spherical radius of 80 inches and having an air weight of 342.57 pounds.
Afterbody	Radiation shield with corrugated shingles of 0.016" René 41 Nickel-Steel alloy over Thermoflex insulation
Recovery Compartment	Covered by 0.220" aluminum panels over Thermoflex insulation

Entrance and Emergency Egress Hatch	Located in the capsule conical (afterbody) section, trapezoidal in shape, and of similar construction to the capsule basic structure. The hatch is designed for emergency egress in event of a land impact. An explosive assembly is incorporated into the hatch, serving as a means when ignited, of breaking seventy (70) titanium attachment bolts. The explosive assembly, consisting of Mild Detonating Fuse (MDF), is mounted about the hatch perimeter, detonated by a push-button initiator, and rendered safed by a single knurled cover and safety pin. The hatch initiator can also be activated from the capsule exterior via a 42 inch long wire lanyard, after the removal of one shingle. Two cabin pressurization tire-type (i.e., schraeder) valves are located in the hatch to permit ground leakage seal checks prior to launch.
Exit Hatch	Inward opening and mounted in small afterbody pressure bulkhead. The hatch is dish shaped, held in place by a retaining ring, and of reinforced titanium and aluminum construction.
Observation Window	Located in the afterbody conical section and consisting of both inner and outer window assembles. The outer window assembly consists of a single pane of 0.350 inch Dow Corning Vycor glass contoured to the capsule's structural shell curvature. The inner window assembly is constructed of three flat panes of glass of trapezoidal shape having an optical fidelity of grade 2N. The two inner panes are 0.340 inch tempered glass and the outermost pane is 0.170 inch Vycor glass. The outer pane contains lateral reference sight lines on the inner and outer surfaces as required by the window mounting angle and fixed optical reference point. The set of lines near the base of the trapezoidal pane provide an eye level sight reference for viewing the horizon compatible with the capsule retrograde attitude of thirty-four (34) degrees with heat sink up. The second set of lines provide an eye level reference for viewing the horizon compatible with a capsule orbital attitude of fourteen and one-half (14.5) degrees from horizontal with heat sink up. The inner surface of the outer window pane and both surfaces of the inner window assembly panes are coated with a single layer of magnesium fluoride (M_gF_2) film for impeding thermal radiation in the cabin.

Life Support Systems	
General	Gaseous oxygen based Environmental Control System (ECS) using high-pressure oxygen bottles to supply breathing gas (100% oxygen) and a Lithium hydroxide (LiOH) absorbent and activated charcoal scrubber to remove carbon dioxide and odors from the breathing gas circuit. Humidity is controlled in the pressure suit heat exchanger using a sponge which is periodically compressed to remove moisture. A solids trap removes any foreign matter such as food particles, hair, nasal excretion, etc. (the trap also incorporates a relief valve in case the system is blocked by solid matter).
Oxygen Supply	Two (primary and secondary) titanium high pressure gas spheres certified to 7,500 psi safe working pressure. Integrated with a 7500 to 100 psig pressure reducing valve.
Suit Compressors	Primary and standby, electric driven
Cabin and Suit Pressures	nominal 5 psia cabin / ambient
Coolant (prelaunch)	Freon 114
Coolant (orbital)	Water based system fed under oxygen pressure from the capsule water tank to both cabin and suit heat exchangers.
Cabin Heating	Three equipment heat exchangers (2 main and one standby) using cabin equipment blower to circulate the atmosphere and maintain both cabin and suit temperature at approximately 80 degrees F.

Stabilization Control Systems	
General	Consists of both the automatic and rate stabilization and control system, horizon scanners, and the reaction control system.
Automatic Stabilization Control System (ASCS)	Provides automatic stabilization and orientation of the capsule from time of separation until landing parachute deployment. Supplies output signals for display, recording, and telemetering of three-axis attitude data, a discrete signal at 0.05 g longitudinal deceleration during re-entry, and attitude signal sector for use in capsule retrograde firing interlock circuit. Related equipment is the horizon scanners, reaction controls, communications system telemetry, attitude displays.
Rate Stabilization and Control System (RSCS)	Operates independent of the ASCS (except for sensing 0.05 g) providing a redundant rate damping feature as a back-up to ASCS damping as necessary. System is integrated with a three-axis hand controller (joystick) providing angular rates approximately proportional to stick deflection. Includes a rate damper, three hand controller potentiometers, fuel subsystem, 6 solenoid control valves, fuel selector valve, and roll and yaw rate transducers. Pitch and yaw rates are dampened to "0" degrees per second by the RSCS when in "Rate Command Mode."
Horizon Scanner	Consists of two AC powered scanner units, one aligned to the capsule pitch axis and the other aligned to the roll axis. They are body mounted inside of the antenna fairing structure providing a conical scan of the horizon via a rotating prism (30 rps) positioned ahead of the scanner lens.
Reaction Control System (RCS)	Automatic and manual control subsystems providing control of the capsule in the roll, pitch, and yaw axes. The system is a pressure-fed, monopropellant / catalyst bed design incorporating right angle firing nozzles producing thrust via the decomposition of hydrogen peroxide (H_2O_2).
Fuel Supply	High strength hydrogen peroxide contained within two flexible torus bladders externally pressurized by 460 psi helium. The hydrogen peroxide fuel can be dumped after main parachute deployment.
Helium Tanks	Two spherical tanks fabricated from fiberglass having a design working pressure of 3,000 psi.
Thrust Chambers	ASCS system uses 12 thrust chambers (6 @ 1 lb. thrust, 2 at 6 lbs. thrust, and 4 @ 24 lbs. thrust) using platinum screens to decompose hydrogen peroxide fuel. RSCS uses six thrust chambers (2 @ 1 to 6 lbs. thrust and 4 @ 4 to 24 lbs. thrust).

Communications Systems	
C-Band Beacon	Avion 152A200-4Q radar tracking beacon (i.e., transponder) compatible with the FPS-16 radar system. Transistorized receiver operating at 5480.00 megacycles and a transmitter (except for its magnetron) operating at 5555.00 megacycles. Power output was 375 watts with a nominal range of 700 nautical miles.
S-Band Beacon	Avion 152A700-6Q radar tracking beacon compatible with SCR-584 Mod. II Radar and VERLORT long range radar. Transistorized receiver frequency operates at 2900.00 megacycles and transmitter operates at 2950.00 megacycles. Power output is 1000 watts with a nominal range of 700 nautical miles.
Two Way HF-UHF Orbital Voice Communications	Two-way amplitude modulated HF transmitter-receiver operating on 15.016 megacycles with a 5.0 watt power output and 10 db signal/noise ratio. UHF transmitter-receivers (primary and backup) operating on 296.8 megacycles with a 2.0 watt (high power) and 0.5 watt power (low power) output and 10 db signal/noise ratio.

Audio Box	Voice-controlled transmit-receive (VOX) to activate transmitters and receivers with a field adjustable threshold level (automatically energized after capsule separation).
Command Receivers	Two frequency modulated transistorized command receivers (similar to AN/DRW-13 receivers) with a total of 20 decoder outputs consisting of 10 channels in each of the receivers and 10 channels in each of the decoders provided. Each command receiver operates at a frequency of 414.0 megacycles and is compatible with FRW-2 ground command transmitters.
Telemetry	HF and LF telemetry transmitters and power supplies. Data is telemetered to ground stations providing real-time information concerning the astronaut, capsule, and life support. HF transmitter operates continuously at 259.7 megacycles at 3.3 watts output power and the LF transmitter operates at 225.7 megacycles at 3.3 watts. The LF unit transmits scientific and aeromedical data by means of four IRIG standard FM subcarriers, one containing PAM modulation (10.5 kc subcarrier) which provides 88 data samples plus two sync pulses for reference, each measured 1.25 times/second.
C and S Band Antennas	Both C and S band antennas consisting of three flush helices for each of the two beacons for omnidirectional coverage with a power divider for each of the two beacons and matched cabling from the power dividers to the antennas.
Biconical Antenna	Operates during prelaunch, launch, orbit, and re-entry phases of the mission, is incorporated into the antenna housing, and is jettisoned at 10,000 feet altitude along with the fairing. The biconical antenna is used with HF and UHF orbital voice communications, both UHF command receivers, and both telemetry transmitters.
UHF Descent Antenna Array	Wire butterfly type descent antenna for omnidirectional coverage permitting simultaneous operation of both telemetry transmitters, UHF back-up voice communications, UHF rescue beacon, and UHF command receivers. The antenna is spring-loaded and extended after a 16-second time delay after antenna fairing separation using a reefing line cutter which severs a tie down cord.
HF Rescue Antenna System	Telescopic whip-type antenna for use with the HF rescue beacon and rescue voice transmitter-receiver. The antenna is stowed in the recovery compartment and extended (using a gas generator) to a length of 16 feet while operating.
HF Diplexer	The unit is used during the recovery phase to connect the output of the HF portion of the HF / UHF rescue beacon and HF rescue voice transmitter to the HF rescue (whip) antenna.

	Navigation Systems
Periscope	Optical unit capable of being extended during the orbital and landing phases and retracted during the launch phase. The unit provides an optical reference point at Z135.59, TY5.780, and X0.00 station lines, based on the astronaut's eye reference point at Z118.20, TY22.82, and RX1.28. The periscope provides an 8 inch diameter circular display with the image plane inclination at approximately 45 degrees from the Y0.00 axis. Specific periscope controls are: 1. Reticle illumination control knob 2. Altitude knob and indicator 3. Drift knob 4. Sun-Moon index control lever 5. Two-position (high and low) magnification change lever 6. Four-position filter selector lever (clear, yellow, red, and neutral) 7. Manual extension and retraction control lever
Navigational Aid Kit	Contains maps, cards, and a pencil, bound together in book-like fashion. All navigational aid kit functions may be performed with inflated pressure suit gloves.
Stereographic Maps	Polar sterographic maps
Cards	Check, chart, and note cards
Pencil	Reliable mechanical type pencil, pencil holder, and retaining line.

	Instrumentation Systems
Cameras	Milliken DMB-8A Astronaut Observer Camera (3 frames per second with 250 feet of Dupont P931A Cronar based film) with Bell and Howell "Ingenue" 10 mm f1.8 lens; Milliken DRM7A Instrument Observer Camera (6 frames per second with 500 feet of Dupont P931A Cronar based film) with Bell and Howell "Ingenue" 10 mm f1.8 lens; Note: Both cameras operated continuously during flight and for 10 minutes after impact.
Tape Recorder	Consolidated Electrodynamics (CEC) tape recorder operates continuously during all phases of the mission and for 10 minutes after impact for recording of astronaut comments, observations, and all voice messages. The recorder has seven heads for recording at a tape speed of 1-7/8 ips and a tape capacity of 4,800 feet using ½-inch Mylar base (3M No. 197) tape.
Commutated Data Recording	Two PDM / PAM commutator / keyer systems signaling inputs at a rate of 112½ samples per second providing 90 data samples, each measured 1¼ times per second, producing a signal wave train.
Cosmic Ray Film Pack	Four photographic recorders of cosmic ray collisions.
Voltage Controlled Subcarrier Oscillators	Voltage controlled subcarrier oscillators receiving PAM outputs from the commutators, 3 vdc, and pitch, roll and yaw signals
Compensating Oscillator	Fixed frequency compensating oscillator for monitoring wow and flutter adjusted to operate at 3,125 cps with an adjustable voltage output.
Mixer Amplifier	Two amplifiers using zener diode power supplies converting 24 vdc to 6 vdc for use by the subcarrier oscillators. One mixer mixes and amplifies oscillator outputs to telemetry transmitter "B" only while the other mixes and amplifies oscillator outputs to telemetry transmitter "A" (reducing the 10.5 KCPS mixer output to a level compatible with the tape recorder inputs).

	Survival Gear
Survival Kit	A partitioned container with one compartment containing a modified PK-2 one-man raft and signal mirror with the other compartment holding: 1. Chemical desalting kit (for 8 pints of water) 2. Dye marker packets 3. Shark chaser packets 4. Battery powered survival light (ACR-4-E or equivalent) 5. First aid kit 6. Signal whistle 7. Small can survival rations 8. Approximately 18 full-size waterproof matches 9. Ten feet multi-braided nylon line 10. Small pocket knife 11. SARAH radio beacon with antenna and battery (Ultra RB-5/7)
Personal Parachute	A personal astronaut parachute is provided in the event of the failure of the capsule main and reserve parachutes. It is mounted inside and below the explosive egress hatch sill and must be manually connected to the astronaut's safety harness before use.

The attached CDROM includes NASA and contractor drawings for the Mercury Capsule and Redstone booster.

— Appendix C —
Locations of Mercury Spacecraft / Test Vehicles

No.	Mission	Location
1	Beach abort test	New York Hall of Science, Corona Park, NY
2	Mercury-Redstone 1	KSC Visitors Center, Kennedy Space Center, FL
3	Little Joe 5	Destroyed during launch from Wallops Island, VA
4	Mercury-Atlas 1	Destroyed during launch, wreckage at Kansas Cosmosphere and Space Center, Hutchinson, KS
5	Mercury-Redstone 2	California Museum of Science and Industry, Los Angeles, CA
6	Mercury-Atlas 2	Houston Museum of Natural Science, Houston, TX
7	Mercury-Redstone 3 (Freedom 7)	US Naval Academy, Annapolis, MD
8	Mercury-Atlas 3/4	Dismantled in 1976 by NASM for spare parts
9	Mercury-Atlas 5	North Carolina Museum of Life and Science, Durham, NC
10	Unflown	Kansas Cosmosphere and Space Center, Hutchinson, KS
11	Mercury-Redstone 4 (Liberty Bell 7)	Recovered July 20, 1999; property of KCSC; on tour of United States
12B	Mercury-Atlas 6 Backup (Unflown)	Aviodrome-Schipol, Amsterdam, the Netherlands
13	Mercury-Atlas 6 (Friendship 7)	National Air and Space Museum, Washington, DC
14	Little Joe 5A/B	Virginia Air and Space Center, Hampton, VA
15B	Unflown	Ames Research Center, Moffett Field, CA
16	Mercury-Atlas 8 (Sigma 7)	United States Astronaut Hall of Fame, Titusville, FL
17	Unflown	US Air Force Museum, Wright-Patterson AFB, Dayton, OH
18	Mercury-Atlas 7 (Aurora 7)	Museum of Science and Industry, Chicago, IL
19	Mercury-Atlas 8 Backup (Unflown)	Communication, Lucerne, Switzerland
20	Mercury-Atlas 9 (Faith 7)	Space Center Houston, Houston, TX

— Appendix D —
Project Mercury Contractors

The following is a partial list of the subcontractors working under the McDonnell Aircraft Corporation on Project Mercury; in reality, there were over 4,000 companies supporting the McDonnell Aircraft Corporation during Project Mercury.

- McDonnell Aircraft Corporation, St. Louis, Missouri – Prime contractor.
- Convair Astronautics Division, San Diego, California – Atlas launch vehicle.
- Chrysler Corporation Missile Division, Detroit, Michigan – Redstone launch vehicle.
- North American Aviation Inc., El Segundo, California – Little Joe launch vehicle airframe.
- Ventura Division (formerly Radioplane) of the Northrop Corporation – Spacecraft landing and recovery system.
- B.F. Goodrich Company, Akron, Ohio – Mercury spacecraft astronaut pressure suit.
- Western Electric Company, New York, New York – Mercury world-wide tracking network.
- Minneapolis-Honeywell Regulator Company, Minneapolis, Minnesota – Stabilization system for the Mercury spacecraft.
- Bell Aerospace Corporation, Buffalo, New York – Reaction control system.
- AiResearch Manufacturing Division of the Garrett Corporation, Los Angeles, California – Environmental control system.
- Perkin-Elmer Corporation, Norwalk, Connecticut – Mercury spacecraft periscope.
- Barnes Engineering Company, Stamford, Connecticut – Horizon scanners.
- Atlantic Research Corporation, Alexandria, Virginia – Escape tower jettison rocket and posigrade rockets for Mercury spacecraft.
- Thiokol Chemical Corporation, Elkton, Maryland – Retrograde rockets for Mercury spacecraft.
- Lockheed Propulsion Company, Redlands, California – Rocket motor for the spacecraft escape tower.

- Cincinnati Testing and Research Laboratory of the Studebaker-Packard Corporation – Spacecraft heat shield.
- Aeronca Corporation, Middletown, Ohio – Honeycomb panels for spacecraft impact landing system.
- Collins Radio Corporation, Cedar Rapids, Iowa – HF and UHF voice communications and UHF recovery antenna for spacecraft.
- Motorola, Corporation, Franklin Park, Illinois – Command receivers for spacecraft.
- Texas Instruments, Corporation, Dallas Texas – Onboard telemetry communications.
- Melpar, Corporation, Falls Church, Virginia – C and S-band antennas.
- Avion Division and G. E. – C and S-band beacons, onboard communications.
- Consolidated Electrodynamics Corporation, St. Louis, Missouri – Eight channel tape recorder for spacecraft.
- D.B. Milliken Company, Arcadia, California – Instrumentation and pilot observer cameras for spacecraft.
- Waltham Precision Instrument Company, Waltham, Massachusetts – Satellite clock (this contract was canceled on December 14, 1960 and McDonnell Aircraft's orbital timing device used as a replacement).
- Bendix Radio Division of the Bendix Corporation, Baltimore, Maryland – Air to ground communications, radar, and acquisition systems for Mercury worldwide tracking network.
- International Business Machines Corporation, New York, New York – Computers and computer programming for Mercury tracking network.
- Stromberg-Carlson (Division of General Dynamics), Rochester, New York – Control center consoles for Mercury worldwide tracking network.
- Grumman Aircraft Engineering Corporation, Bethpage, New York – Operations analysis study of recovery problems associated with a three-orbit mission.
- Philco Corporation, Philadelphia, Pennsylvania – Range monitors for Mercury worldwide tracking network.
- Pan American Airways, Cape Canaveral, Florida – Atlantic Missile Range operations.
- Space Technology Laboratories, Redondo Beach, California – Analysis of flight instrumentation and design trajectories for the Mercury-Atlas program.

— Appendix E —
Recovery Equipment Specifications

Ocean Explorer 6000 Side Scan Sonar:

General	
Depth Rating	6 to 6,000 meters
Tow Cable	Single coaxial cable, 10,000 meters
Towing Configuration	Two-body tow system with neutrally buoyant towfish
Dimensions	4 meters length, 1.5 meters height, and 1.2 meters width
Towfish Weight	2,700 lbs. in air

33/36 kHz Wide Swath Sonar	
Beam width	1.6 degrees Horizontal, 40 degrees Vertical
Transmit Power	190 or 2,580 watts per channel
Bandwidth	0.4 to 6.0 kHz, swath width dependent on manual override
Swath Limits	500 to 5,000 meters nominal in 500 meter increments

120 kHz High Resolution Sonar	
Beam width	1.6 degrees Horizontal, 60 degrees Vertical
Transmit Power	70 to 1,000 watts per channel
Bandwidth	2 to 24 kHz, swath width dependent on manual override
Swath Widths	100, 250, 500 and 1,000 meter nominal swaths

Additional Sensors / Capabilities	
Towfish Sensors	Heading, pitch, roll, pressure (depth), altimeter
Navigation	7-14 kHz responder / transponder, 10 km range
Swath Bathymetry	Isophase interferometry over a 1 km swath
Sub-Bottom Profiler	4.5 kHz, 500 watts transmit power
Emergency Release	Acoustically triggered emergency surfacing

Image Display System	
Image Display	Q-MIPS 1280 x 1024 high resolution with 256 colors
Display Format	Slant-range and speed-corrected, beam / grazing angle compensation
Data logging Format	1.2 GB Optical Disks
Gray Scale Printer	EPC 9800 dual-channel recorders, analog and digital
Color Printer	Tektronix RGB III, 300 dpi resolution, 4096 colors

Note: All equipment specifications are as fitted for the Liberty Bell 7 expedition

Magellan 725 Remotely Operated Vehicle:

General	
Depth Rating	7,000 meters
Umbilical	Optical fiber cable (3 fibers), 9150 meters
Dimensions	2.4 meters length, 1.5 meters height, and 1.4 meters width
Weight	4,500 lbs. in air
Propulsion	Two axial, three vertical, and two lateral Innerspace thrusters
Hydraulic Power	25 hp at 3,000 psi

Instrumentation	
Scanning Sonar	Ametek 250A CTFM sonar, 107-122 kHz, 1-610 meters range
Video Cameras	SIT low light b/w camera with wide angle lens; High resolution CCD color zoom camera; Broadcast quality 3-chip color camera; WHOI Sony wide screen format, high resolution, broadcast quality color camera mounted in titanium housing on hydraulic tilt unit
Video Recording	Two Hi-8 video cassette recorders; Two Sony digital beta video recorders
Photographic	35 mm survey cameras, 250 frames; 35 mm survey camera, 750 frames
Lighting	Four 250 watt lights; Single head strobe light, 150 watts; Two Deepsea Power & Light 450 watt HMI lights
Pan / Tilt	Hydraulic pan / tilt unit with titanium shaft, 340 degrees Pan and 160 degrees Tilt
Depth Sensor	Paro Scientific pressure transducer, 0.05% accuracy
Heading Sensor	Fluxgate compass, ±1 degree Accuracy
Altitude Sensor	High frequency altimeter, 200 kHz, 30 meter range
Auto Controls	Auto depth, auto altitude, auto heading
Electrical cabling / connectors	Pressure Balanced Oil Filled (PBOF) and neoprene coated electrical wiring, stainless-steel electrical connectors (Impulse manuf.), teflon coated copper wiring throughout

Recovery Tools / Capabilities		
Manipulators		Seven function, rate controlled, 200 lbs. lift; Five function, rate controlled, 200 lbs. lift
ROV / Depressor Payload		Maximum 1,200 lbs. payload

Recovery Line Spooler
Aluminum frame structure fitted with spooler holding approximately 25,000 feet of 3/8 inch Yale Cordage Aracom Miniline kevlar recovery line (average break strength of 12,800 lbs.), terminating in approximately 100 feet of 1 inch kevlar line (average break strength of 100,000 lbs.), for lifting through the air / sea interface.

Capsule Recovery Tools
Four clamp style mechanically actuated recovery tools, with jaws machined to fit around the exterior / interior of Mercury spacecraft escape tower mating ring. Each tool, which is of mild steel welded construction, uses screw threads to open and close, is fitted with a lifting bail, and has a standard ROV T-Handle gripper interface.

Lost Spacecraft

Target	Depth	Date	Location	Organization	Asset	Comments
Submarine F-4 (USS Skate)	305	1915	off Barber's Pnt, Hawaii	US Navy	Divers	One of the first uses of recompression chamber
Submarine S-51 (SS-162)	152	1926	Block Island, NY	US Navy	Divers	Loss of 23 crew
Submarine S-4 (SS-109)	110	17 Mar 1928	Cape Cod	US Navy	Divers	Loss of 40 crew
P&O Liner Egypt	396	1929	Bay of Biscay	SORIMA Co.	Articulated Suits	Gold, silver, and coinage recovered
USS Squalus (SS-192)	243	1939	off Portsmouth, NH	US Navy	Divers	First use of submarine rescue chamber
SS Niagara	470	Dec 1941	Hauraki Gulf, NZ	United Salvage Proprietary	Diving Bell	Gold bullion recovered
USS Thresher (SS-593)	8,500	Oct 1963	off New England	US Navy	Trieste I	Loss of 129 crew
Hydrogen Bomb	2,550	7 Apr 1966	near Palomares, Spain	US Navy	CURV III	Lost during mid-air collision
Alvin DSV	5,500	28 Aug 1969	off Cape Cod	US Navy	Aluminaut DSV	Lost during launching operations
Pisces III	1,575	1 Sept 1973	off Cork, Ireland	Vickers Oceanics/US Navy	Pisces II/CURV	Deepest submarine rescue
Submarine K-129	17,000	Aug 1974	1,700 nm NW Hawaii	Summa Corp/CIA/	Clementine ROV	AKA Project Jennifer
HMS Edinburgh	800	5 Oct 1981	Barents Sea	Wharton Williams	Divers	Sunken WW II Royal Navy Cruiser
Air India Flt. 182	6,200	Jul-Dec 1985	off Cork, Ireland	Eastport Int/US Navy	SCARAB II	Worst air disaster at sea
Space Shuttle Challenger	1,300	Feb-Aug 1986	off Cape Canaveral	Eastport Int/US Navy	Gemini	Operations in 3 to 4 knots of surface current
SS Central America	9,500	1986	Atlantic, off Georgia	Columbus America Dis. Grp.	Nemo ROV	Used unique small object recovery techniques
SAA Flt. 295	14,600	Jan-Mar 1989	near Mauritius	Eastport Int/US Navy	Gemini 6000 ROV	Used upgraded Gemini ROV
UAL Flt. 811	14,200	Sep-Oct 1990	off Hawaii	Oceaneering/US Navy	Sea Cliff	Cargo door recovery
CH-46 Helicopter	17,251	1992	off Wake Island	Oceaneering Int/US Navy	CURV III	Deepest ocean salvage
SS John Barry	8,500	Nov 1994	Arabian Sea	Ocean Group	Drillship Flex LD	Sunken WWII US Liberty ship
Liberty Bell 7	16,043	20 Jul 1999	Atlantic Ocean	Oceaneering/DCI/LB7	Ocean Dis. ROV	Deepest commercial salvage
CSS Hunley	28	8 Aug 2000	Charleston Harbor	Oceaneering/Nat. Park Service	Barge Karlissa B	Sub was used on first successful submarine attack
SSGN Kursk (K-141)	354	8 Oct 2001	Barents Sea	Smit Tak/Mammoet	Giant 4	Raised large Oscar II nuclear submarine
MV Ehime Maru	1,968	11 Oct 2001	off Honolulu, Hawaii	Smit Tak/US Navy	MV Rockwater II	Deep water, large object recovery

Note: All depths are in feet of sea water (fsw).

— Appendix F —
Notable Object Recovery Operations

See table on page 208.

— Appendix G —
Bibliography

NASA Government Documents:
- *Project Mercury Postlaunch Trajectory Report for Mercury-Redstone Mission 4*, NASA Space Task Group, Langley AFB, Virginia, October 16, 1961
- *Post Flight Debriefing of Virgil I. Grissom*, Memorandum for Associate Director, NASA – Manned Spacecraft Center, Langley Field, Virginia, July 21, 1961
- *Astronaut Recovery Handbook*, McDonnell Aircraft Corporation, (Capsules 11 & 15), SEDR 189-11 & 15, Serial No. 37, June 1, 1961
- *Project Mercury Mission Directive for Mercury-Redstone No. 4 (Capsule No. 11)*, NASA Space Task Group, Langley Field, Virginia, July 7, 1961
- *Project Mercury Technical Summary of Mercury-Redstone No. 4*, NASA Space Task Group, Langley Field, Virginia, July 11, 1961
- *Project Mercury Mercury-Redstone No. 4 Recovery Requirements*, NASA Space Task Group, (undated)
- *Notes on Explosively Actuated Side Hatch*, (undated)
- *Mercury Capsule No. 11 Configuration Specification (Mercury-Redstone No. 4)*, McDonnell Aircraft Corporation, March 6, 1961
- *Project Mercury Familiarization Manual*, NASA Manned Satellite Capsule, McDonnell Aircraft, November 1, 1961
- Ledford, Harold, *Actual Trajectory of Mercury-Redstone Flight test MR-4 (U)*, NASA George C. Marshall Space Flight Center, Huntsville, Alabama, September 22, 1961
- *Index and Test Results, Part 1 of the Firing Test Report Mercury-Redstone Vehicle MR-4 (U)*, AMR Test Nr. 1809, NASA George C. Marshall Space Flight Center, Reports and Publications Section, August 15, 1961
- Hinds, Noble F., *Instrumentation Operations Analysis, Part IIb of the Firing Test Report Mercury-Redstone MR-4 (U)*, George C. Marshall Space Flight Center, August 24, 1961

- Clarke, W.G., *Preliminary Evaluation of Mercury-Redstone Launch MR-4 (U)*, George C. Marshall Space Flight Center, August 22, 1961
- *Project Mercury Memorandum Report for Mercury-Redstone No. 4 (MR-4)*, NASA Space Task Group, Cape Canaveral, Florida, August 6, 1961
- *Results of the Second U.S. Manned Suborbital Space Flight*, Manned Spacecraft Center, NASA, July 21, 1961

United States Navy Documents:
- Decklog, United States Navy, USS *Conway* (DDE 507), July 16 - 22, 1961
- Decklog, United States Navy, USS *Cony* (DDE 508), July 16 - 22, 1961
- Decklog, United States Navy, USS *Lowry* (DD 770), July 16 - 22, 1961
- Decklog, United States Navy, USS *Randolph* (CVS 15), July 18 - 24, 1961
- Decklog, United States Navy, USS *Stormes* (DD 780), July 16 - 22, 1961
- Decklog, United States Navy, USS *Waller* (DDE 466), July 16 - 22, 1961

Other Documentation:
- *Historic Aircraft*, Dryden Flight Research Center, URL: www.dfrc.nasa.gov/history/x-planes.html, March 13, 2001
- *Major Milestones in Aerospace History at Edwards AFB*, Air Force Flight Test Center, URL: www.edwards.af.mil/history/docs_html/center/major_milestones.html, Jan. 7, 2002
- *X-15 Hypersonic Research Program*, NASA Facts, Dryden Flight Research Center, URL: www.dfrc.nasa.gov/PAO/PAIS/HTML/FS-052-DFRC.html, December 1998
- Stockett, Stewart J., BSChE, Senior Engineer, Materials and Process Development Dept., McDonnell Aircraft Corp., *Metals and Processing Methods Used in the Mercury Spacecraft* (version abridged in 1962 due to government security concerns), Metal Progress Magazine, June 1962

Books:
- *This New Ocean*, A History of Project Mercury, NASA, Scientific and Technical Information Division, 1966
- *Project Mercury*, A Chronology, NASA, Office of Scientific and Technical Information, 1963
- Carpenter, M. Scott, Cooper, L. Gordon Jr., Glenn, John H., Grissom, Virgil I., Schirra, Walter M. Jr., Shepard, Alan B. Jr., Slayton, Donald K., *We Seven*, Simon and Schuster, New York, 1962

- Grissom, Virgil I. "Gus," *Gemini*, The Macmillan Company, New York, 1968
- Lewis, Richard S., *Appointment on the Moon*, Ballantine Books, New York, 1968
- Chapell, Carl L., *Seven Minus One, The Story of Gus Grissom*, New Frontier Publications, Mitchell, Indiana, 1968
- Yeager, Chuck, *Yeager*, Bantam Books, New York, 1985
- Collins, Michael, *Liftoff – The Story of America's Adventure into Space*, Grove Press, New York, 1988
- Grissom, Betty, and Still, Henry, *Starfall*, Thomas Y. Crowell Co., New York, 1974
- Cooper, Gordon, *Leap of Faith – An Astronaut's Journey into the Unknown*, Harper Collins, New York, 2000
- Shepard, Alan, Slayton, Deke, Barbree, Jay, and Benedict, Howard, *Moonshot – The Inside Story of America's Race to the Moon*, Turner Publishing, Inc., Atlanta, 1994
- Slayton, Deke, and Cassutt, Michael, *DEKE! U.S. Manned Space: From Project Mercury to the Shuttle*, Forge Books, New York, 1994
- Cunningham, Walter, *The All American Boys*, Macmillan Publishing, New York, 1977
- Kranz, Gene, *Failure is not an Option*, Simon & Shuster, New York, 2000
- Kraft, Christopher, *Flight – My Life in Mission Control*, Dutton, New York, 2001
- Ley, Willy, *Rockets, Missiles, and Space Travel*, Viking Press, New York, 1959
- Broad, William J., *The Universe Below*, Touchstone, New York, 1998

Telephone and Personal Interviews:
- Telephone Interview, Dr. Max Faget (Mercury Spacecraft Designer), 1987
- Telephone Interview, Capt. H.E. Cook (former Captain, USS *Randolph*, USN Ret.), 1987
- Interview, Capt. Robert H. Gillock (former Navigator, USS *Randolph*, USN Ret.), 1987
- Interview, Dr. James L. Lewis (former helicopter pilot for prime recovery force Hunt Club-1, USMC Ret.), 1987
- Interview, Walter M. Schirra, (former Mercury astronaut, USN Ret.), 1987
- Interview, Robert F. Thompson (former Head of Project Mercury Recovery Operations), 1987
- Interview, John Yardley (former McDonnell Douglas Aircraft Chief Engineer for the Mercury spacecraft), 1987
- Interview, Lowell Grissom (Grissom's younger brother), 2001
- Interview, William Head (Grissom's childhood friend), 2001

Audio and Video Tape:
- *The Flight of Liberty Bell*, NBC Television Coverage, July 21, 1961
- *The Flight of Liberty Bell*, Television Pool Coverage, NBC Television Coverage, July 21, 1961
- Tape Recording of Air-to-Ground Communications, Mercury-Redstone Flight No. 4, NASA, Kennedy Space Center, July 21, 1961
- *In Search of Liberty Bell 7*, Documentary Film, The Discovery Channel, 1999

CDROM

The attached CDROM features
NASA and manufacturer technical drawings
of the Liberty Bell 7 spacecraft and
recovery mission planning.
Also included is the Operations Log
of the Liberty Bell 7 recovery project.

The disc is rated for Windows 95 and higher.
An adequate web browser is also necessary to play the disc.